Introduction to Statistical Methods

Revised Printing
N. Clayton Silver

Kendall Hunt
publishing company

Kendall Hunt
publishing company

www.kendallhunt.com
Send all inquiries to:
4050 Westmark Drive
Dubuque, IA 52004-1840

Contents

CHAPTER 1

Introduction and Descriptive Statistics

INTRODUCTION

Welcome to the wonderful world of statistics. When you think of statistics, of what do you think? Is it the number of people who live in a state? Is it the number of deaths on a highway for a given year? Is it the number of home runs hit per at bat? Whatever it is, there is a common denominator among these questions; that is, it is a number or quantity. Of course, these numbers can determine policies, salaries, and other major decisions. Moreover, have you wondered why your department (i.e., psychology, anthropology, hotel administration) makes you take this course? After all, statistics is nothing more than mathematics, which, of course, has nothing to do with your major, right? Besides, I'm sure you are saying to yourself "I'll never use this stuff, just let me survive this course." It turns out that statistics are tools. If they are tools, then think of me as "Home Depot or Lowe's." That is, I'll give you some tools, show you how to use them, and explain why you use them. With these tools, you can answer questions and tell stories. For example, if Pfizer has developed a drug for Ebola, then does that drug produce significantly fewer symptoms over time than any drug that is out there now, or no drug at all? For a hotel administration major (given that we live in Vegas), might Frank Fertitta be interested in knowing if there is a statistically significant difference in quality of stay (on a scale of 1–5 with 1 being poor and 5 being excellent) between Sunset Station and Palace Station hotels across each of 12 months for a particular year? Although I am definitely not an anthropologist, perhaps a question that can be raised is the following: Is there a statistically significant difference in the head measurements of Nama, Kung (both African cultures), Trumai, and Timbira (both South American cultures) adult males? Being a sports aficionado, one of the questions that my former undergraduate student and I raised was the following: Is there a statistically significant difference in the average salaries of the offensive pro football players namely, receivers, quarterbacks, and linemen? Furthermore, did they differ across conferences (AFC and NFC) and divisions (East, North, Central, and West)? As you are able to gather, there are a plethora of important questions that statistics will allow you to answer. Hopefully, you'll find telling stories and finding answers to these and many other questions to be an exciting proposition.

So, let's begin addressing some of the basic terminology. First, it is always nice to know what you are studying, namely, statistics. A **statistic** is a quantity calculated from a sample. It is designated in Latin (or alphabetic) letters. For example, $\overline{X}$ (x-bar) is the sample mean, s^2 represents the sample variance, and s represents the sample standard deviation. Of course, we'll discuss how to calculate these a bit later. Yet, a **parameter** is a quantity calculated from a population. These are designated in Greek letters. For example, μ (mu, pronounced "mew") is the population mean, σ^2 (sigma squared) is the population variance, and σ (sigma) represents the population standard deviation. A **population** is a set of all objects that we are interested in researching. For example, we tend to think of populations as extremely large numbers, such as the population of Las Vegas (roughly 2 million) or the population of Harmony, California (18 people, I believe). You folks are a population of students taking my class at a certain time, room, and semester. If I create a specific religion of which I am the only participant and observant, then I would be a population of 1. A **sample**, by comparison, is a subset of the population. For example, if I take everyone sitting in the first row of the class, then that would be a sample. In psychology, we strive to obtain a random sample in order to generalize our results to the population. A **random sample** is when each member of the population has an equal likelihood of being selected. For example, in the lottery, each of the 6 numbers out of 49 (the population of numbers) would have an equal probability of being selected. However, dealing with humans may be a different story. For example, when you read research articles in psychology from various universities such as Harvard, Stanford, Dartmouth, or even University of Nevada-Las Vegas (UNLV), they will "randomly" sample a certain number of students and then try to generalize their results to the general population. Is this a random sample? Do you think that one can get a random sample in Las Vegas in order to generalize to the population? For example, could one get a random sample at the Strip? The Strip is usually full of tourists; do they really represent the general population? They certainly spent money to come here and many of the "locals" avoid the Strip, unless they are playing tourist guide. So this might not be the best possibility. Maybe a better possibility is Walmart. Walmart attracts about everyone; this transcends across ethnicities, religions, sexes, socioeconomic statuses, age, etc. Another possibility would be to randomly sample at supermarkets. However, the problem here is that you would need to sample across areas; that is, Summerlin and Green Valley may have a certain demographic structure, whereas other areas such as North Las Vegas and East Las Vegas have another. Therefore, you would want to sample different demographic areas in order to obtain a reasonable random sample. Although it may be theoretically feasible, it may not be economically possible to do so. Although obtaining a random sample may be a panacea, it is still possible to get pretty darn close, even in Las Vegas.

After we obtain a random sample and have these volunteers participate in our experiment, we will then obtain data. When we analyze the data, we might find a statistically significant difference between groups (e.g., males and females) with regard to what we are measuring (e.g., a general anger scale). Therefore, what do we mean when we use the word significance? In English, significance means importance. Unfortunately, that definition does not quite fit statistically. In statistics, **significance** means unlikely to have occurred by chance alone. For example, suppose when we perform our statistical mumbo jumbo, we find that males have a significantly higher score on the anger scale than did females. In psychology, we use 95% or 99% significance levels. If it were significant at the .05 level, then we would say that there was a 95% probability that this result did not occur by chance alone. I like to think of statistical significance as the "Ivory Soap" phenomenon. In the old days, there was a commercial on television stating that Ivory Soap was 99 and 44/100% pure; that is, "significant" soap. Of course, 56/100% is error, which would include other materials such as dyes and perfumes. Just because a finding is statistically significant, doesn't necessarily mean it is important. For example, suppose 50 people drink a sip of Coca Cola followed by a sip of Pepsi and rate each in terms of taste and another 50 people drink a sip of Pepsi followed by a sip of Coca Cola and rate each in terms of taste. Moreover, if we find a statistically significant difference with regard to the order effect, then what does that mean? Is that really important or is it just a bad sampling break? Conversely, you might obtain a finding that is not statistically significant (what we call nonsignificant) and yet it could still be important. For example, suppose that we find no statistically significant differences among a control group (no drug), experimental drug 1, and experimental drug 2 for reducing the symptomatology of a new strain of flu. The experimental drugs don't seem to work, so they should not be placed on the market. This is certainly an important finding to the drug manufacturers, the Food and Drug Administration (FDA), and the public (especially those who are afflicted with the new strain of flu).

In this course, we will examine two different types of statistics. **Descriptive statistics** are numbers that summarize or describe data and **inferential statistics** allow us to test hypotheses about the differences between groups on the variable being measured. We begin by examining descriptive statistics, specifically the measures of central tendency, measures of dispersion, z-scores, and confidence intervals (the latter two can also be inferential statistics, but we are treating them as descriptive here).

Measures of Central Tendency

Three of the more common descriptive statistics that you will find in the statistical software packages are the mean, median, and mode. The mean is commonly used as a descriptive statistic in the journals.

1. **Mean**. The mean is the arithmetic average. For those of you who don't get teed off while you are golfing, let's suppose that you play three rounds of 18 holes and scored the following:

   Round 1 = 82

   Round 2 = 78

   Round 3 = 89

   In order to obtain the mean,

   Step 1: Add up all the scores 82 + 78 + 89 = 249

   Step 2: Divide by the number of pieces of data = 3 in this case

   Therefore, the mean = 249/3 = 83.

   Now, let's put that into some statistical notation $\overline{X} = \Sigma X/N$; in which $\overline{X}$ is the sample mean; if it were μ, then it would be the population mean (both are calculated in the same way). X is the raw score and Σ is sigma, which means to add up the raw scores. N stands for the number of pieces of data. In numerical terms, $\Sigma X = 249$ and $N = 3$; therefore, the mean = 83. The only difference between the sample mean and the population mean is in terms of context. For example, if these are the only rounds that I played at TPC, then this would be the population mean. Whereas, if these were three rounds from a group of rounds that I played that week at various golf courses, then this would be a sample mean. Likewise, if these were rounds from three randomly selected different golfers among a group of 20 golfers, then this would be a sample mean. However, if these were rounds obtained by the only professional golfers from among 20 golfers, then this would be the population mean.

2. **Median**. The median is the middlemost score. For example, suppose you have these five scores from a statistics quiz: 30, 50, 10, 80, and 20.

Step 1: Order the scores from smallest to largest

10, 20, 30, 50, 80

Step 2: Cross out the extremes. In this case, you would cross out 10 and 80 and the next extremes are 20 and 50. This would leave you with 30 (the median).

This will always work when you have an odd number of numbers. In this case, I had five scores. However, suppose you have an even number of numbers, like the following:

30, 50, 10, 80, 20, 40

Step 1: Order the scores from smallest to largest.

10, 20, 30, 40, 50, 80

Step 2: Cross out the extremes. By doing so, you would end up with 30 and 40.

Step 3: Take the mean (or midpoint – same thing here) of the two scores 30 + 40 = 70/2 = 35. Thus, the median here is 35.

3. **Mode**. The mode is the most frequently occurring score. Suppose you have the following scores from 10 participants:

   1, 1, 1, 3, 5, 5, 7, 8, 9, 9

   The mode is 1 (a lonely number) because it occurred three times. If you have the following scenario:

   1, 1, 1, 3, 5, 5, 5, 8, 9, 9

   then there are two modes: 1 and 5. Some of the statistical software packages might state "not unique," meaning that there is more than one mode.

   Think of the measures of central tendency as the 3M company—mean, median, and mode.

Measures of Dispersion

Measures of dispersion are examining the spread of the scores or how far apart the scores are from each other. Let's take a look at a few of these.

1. **Range.** This is the largest score minus the smallest score. So, if the highest score on my statistics test was an 80 and the lowest score was 10, then the range would be 70.

2. **Variance.** This is the average of the squared deviations about (or from) the mean. The variance can be computed from a sample (s^2) or a population (σ^2). Moreover, it can also be computed heuristically (the discovered method, or as I affectionately call it, the hard way) and computationally (it looks more menacing, but it is actually easier to compute).

   Suppose we examine the number of times that rats make correct choices in an Olton maze (i.e., obtaining food) out of 10 possible trials. We will randomly sample five rats from a group of 30. Here were the results:

   Rat 1 = 1

   Rat 2 = 2

Rat 3 = 3

Rat 4 = 4

Rat 5 = 10

These would be our Xs or raw scores. In order to compute the sample variance via the heuristic formula, we would use the following formula:

Heuristic Formula for Sample Variance

X	X − $\overline{X}$	(X − $\overline{X}$)²
1	−3	9
2	−2	4
3	−1	1
4	0	0
10	6	36
$\overline{X} = 4$		$50 = \Sigma(X - \overline{X})^2$

$$\frac{\Sigma(X - \overline{X})^2}{N - 1} = \frac{50}{4} = 12.5 = s^2$$

As you can see in column 1, these would be the raw scores or the number of times the rats made the correct choice. At the bottom of the column is the mean, which was obtained by adding the Xs = 20 and dividing by the number of scores (5), so the sample mean ($\Sigma X/N$) = 4.

In column 2, we subtract out the mean from each raw score. For example, for Rat 1, take the raw score 1 and subtract out the mean (4), thereby leaving us with −3. We do this for each raw score. You'll notice that if you add all these together $\Sigma(X - \overline{X}) = 0$. If this were the formula, then it would mean that there was no variability among the scores, which of course, is preposterous. Hence, that is the reasoning for column 3.

In column 3, we take the result for each rat in column 2 and square it. Hence, for rat 1, column 2 was −3, and when squared it equals 9. Any time you square numbers, you introduce variability into the mix. In this case, you'll notice that by squaring these numbers, we are obtaining the squared deviations from the mean. Hence, rat 5 with a score of 10 will have a larger value (farther away from the mean of 4 than will rat 3, with the score of 3). Of course, rat 4 hit the mean; therefore, there is no variability added. Nevertheless, when we add up all the values from column 3, the sum of the squared deviations from the mean will equal 50.

Finally, we divide the result from column 3 (50) by N − 1 (or 4 in this case). Therefore, the sample variance is 50/4 = 12.5. Let's recap the steps.

Step 1: Find the mean of the raw scores

Step 2: Subtract the mean from each raw score value to create a deviational score

Step 3: Square each of the deviational scores (column 3)

Step 4: Add those up

Step 5: Divide that result by N − 1.

If you have a multitude of numbers, this procedure can be quite laborious. So, that provides the impetus for using the computational formula.

$$s^2 = \dfrac{\sum X^2 - \dfrac{(\sum X)^2}{N}}{N-1}$$

Computational Formula for Sample Variance

X	X²
1	1
2	4
3	9
4	16
10	100
20 = ΣX	130 = ΣX²

The first step would be to add all of the raw scores, which gives us 20. Next, square each of the raw scores and then add those numbers. This gives us a total of 130. Let's put the numbers into the equation:

$$\dfrac{130 - \dfrac{(20)^2}{5}}{4}$$

So, what we have is (20)²/5 = 80. Subtract that from 130, which gives us 50. If we take 50 and divide by 4, then we'll get 12.5. Voila, the same variance is obtained whether you use the heuristic or computational formulas.

Let's suppose that instead of randomly sampling 5 rats from a group of 30, we are interested in dealing only with brown rats. In this case, I have a total of 5 brown rats, thus it would constitute a population. Now, we would be dealing with the population variance (σ^2). The heuristic formula is as follows:

$$\dfrac{\sum (X - \mu)^2}{N}$$

In this formula, you'll notice that we are dealing with μ, or the population mean, rather than the sample mean ($\overline{X}$). Of course, that is because we are computing the population variance. When we computed the sample variance, then we needed the sample mean. The second difference is N is now the denominator as opposed to N − 1. This would get us into biased versus unbiased statistics, which I will not address here.

Heuristic Formula for Population Variance

X	X − μ	(X − μ)²
1	−3	9
2	−2	4
3	−1	1
4	0	0
10	6	36
μ = 4		

$$50 = \Sigma (X - \mu)^2$$

$$\frac{\Sigma (X - \mu)^2}{N} = \frac{50}{5} = 10 = \sigma^2$$

This procedure is extremely similar to that of the sample variance. Once again, the mean is 4 (population mean, this time). But, the mean is computed the same way. In column 2, we subtract out the population mean from each raw score. In column 3, we take the result for each rat in column 2 and square it. Once again, when we add up all the values from column 3, the sum of the squared deviations from the population mean will equal 50.

Finally, we divide the result from column 3 (50) by N (or 5 in this case). Therefore, the population variance is 50/5 = 10. Let's recap the steps.

Step 1: Find the population mean of the raw scores

Step 2: Subtract the population mean from each raw score value to create a deviational score

Step 3: Square each of the deviational scores (column 3)

Step 4: Add those up

Step 5: Divide that result by N.

$$\sigma^2 = \frac{\Sigma X^2 - \frac{(\Sigma X)^2}{N}}{N}$$

Computational Formula for Population Variance

X	X²
1	1
2	4
3	9
4	16
10	100
20 = ΣX	130 = ΣX²

Once again, the procedure is exactly the same as it is for the sample variance. The first step would be to add all of the raw scores, which gives us 20. Next, square each of the raw scores and then add those numbers. This gives us a total of 130. Let's put the numbers into the equation:

$$\frac{130 - \frac{(20)^2}{5}}{5}$$

What we have is $(20)^2 / 5 = 80$. Subtract that from 130, which gives us 50. If we take 50 and divide by 5, then we'll get 10. Once again, the same variance is obtained whether you use the heuristic or computational formulas.

Standard deviation—The third measure of dispersion is the standard deviation. Mathematically, this is the square root of the variance. It can also be thought of as the square root of the square of the average deviations about (from) the mean. In our example, the heuristic formula of the sample standard deviation would be

$$s = \sqrt{\frac{\sum(X - \overline{X})^2}{N - 1}}$$

The computational formula for sample standard deviation would be

$$s = \sqrt{\frac{\sum X^2 - \frac{(\sum X)^2}{N}}{N - 1}}$$

In our example, we would square root 12.5, which would give us a sample standard deviation of 3.5. Likewise, to get the population standard deviation, square root the population variance equation. The heuristic standard deviation formula for the population is

$$\sigma = \sqrt{\frac{\sum(X - \mu)^2}{N}}$$

For the computational formula, the population standard deviation would be

$$\sigma = \sqrt{\frac{\sum X^2 - \frac{(\sum X)^2}{N}}{N}}$$

In our example, the population standard deviation is equal to the square root of 10, which is 3.16. We tend to use the standard deviation more often as a descriptive statistic because the number is smaller and easier to think about. For example, a number of years ago, the Graduate Record Examination (GRE – for those of you who want graduate school) had a mean of 500 for the verbal section and a standard deviation of 100. That means most of the scores fell between 400 and 600. One could also say the verbal section had a mean of 500 and a variance of 10,000 (or 100^2). It is easier to think about 100 than 10,000. Hence, in the scholarly journals, that is why you will see means and standard deviations as the descriptive statistics rather than means and variances.

What does this mean? The mean and standard deviation are important only in terms of the context of the work. For example, suppose someone says that I have a mean of 6 and a standard deviation of 2. If I asked you to rate a chain saw on a scale of 0 (not at all careful) to 8 (extremely careful) with regard to how careful you would be when using it, then with a mean of 6 and a standard deviation of 2, most of the scores would fall between 4 and 8, meaning that this is a reasonably hazardous product and that one would be careful. However, suppose this was a statistics test of 100 points and the mean was 6 and the standard deviation was 2. Again, most of the scores would fall between 4 and 8, telling us that either no one studied or the professor was incoherent and potentially incompetent. No jokes or even insinuations here, thank you. Obviously, the context is all important. Just producing these numbers without context is meaningless.

CLASS EXAMPLE

Show the heuristic and computational formulas for the sample and population variances and standard deviations of the following numbers: 2, 4, 6, 8, 10 (number of free throws made out of 10 by five members of the UNLV basketball team).

Types of Distributions

I'm sure that you know the old adage "a picture is worth a thousand words." So, when describing the data, one can produce a picture to indicate how the data look. In order to draw the picture, we have the X-axis (abscissa), which indicates the score, and the Y-axis (ordinate), which indicates the frequency or number of times that something occurs. For example, suppose you had these scores on my statistics test:

23, 33, 35, 45, 46, 49, 53, 54, 60. This could be depicted in the following manner:

Figure 1.1 Frequency polygon for scores on the statistics test

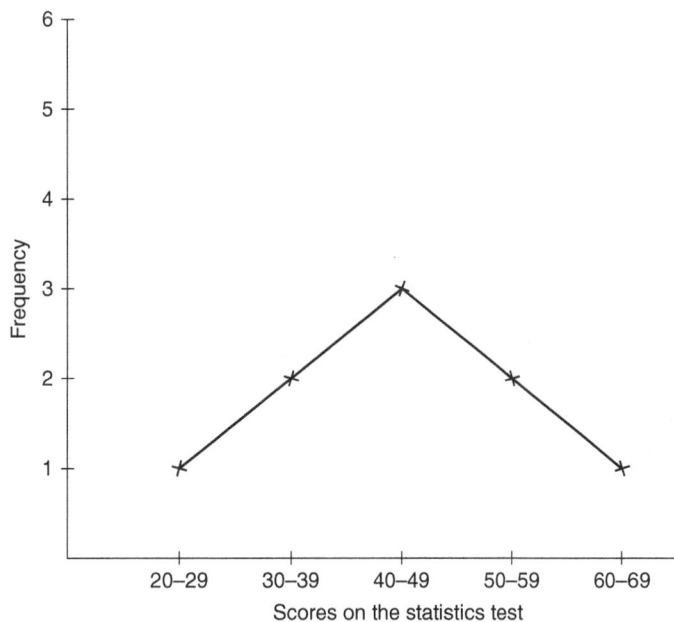

Figure 1.1 demonstrates a frequency polygon. A **frequency polygon** shows the distribution of subjects scoring among various intervals. In this case, because there is such a small sample size, we would use intervals for the X-axis. When you are using intervals, make sure that they are always equal (10–20, 20–30, etc.) rather than 10–25, 26–32, etc. Moreover, don't use each data point on the x-axis. In other words, if you went from 1–60, the best you would find is a frequency of 1. A picture as shown in Figure 1.2 would be meaningless.

Figure 1.2 A meaningless frequency polygon for scores on the statistics test

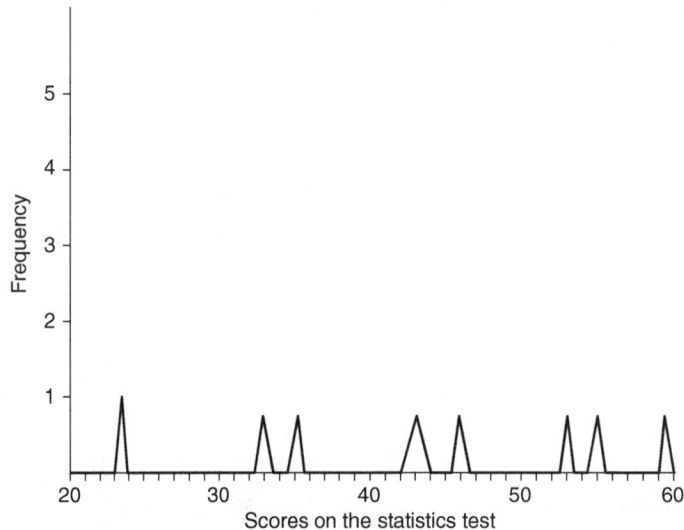

Make sure that you label your axes appropriately and specifically. On the X-axis, in this example, the label is scores on the statistics test, whereas on the Y-axis, it is frequency. If you don't label the axes, then the reader is clueless as to what the numbers represent.

Rather than playing "connect the dots" and drawing a frequency polygon, it is also possible for one to make a histogram of the data. A **histogram** is a graphical representation of the data using bars. These bars are attached to each other given that the x-axis is continuous. Continuous data means that the data are not categorical (i.e., male or female), but rather along some type of continuum (e.g., statistics test scores from 0 to 100). A **bar chart**, however, is when the x-axis is categorical in nature. Consequently, the bars are not attached to each other. In both cases, the y-axis is usually frequency or percentage of responses.

In the graph shown in Figure 1.3, you'll notice symmetry. That is, the left and right sides look alike (if you split the graph down the middle). A distribution like this is called the **normal distribution**. A normal distribution, has many names including the bell curve (as it looks like a bell) or the Gaussian distribution.

In a normal distribution, or in any symmetric distribution, the mean, median, and mode are in similar places (toward the middle of the distribution). Moreover, this type of distribution exhibits kurtosis. **Kurtosis** is the peakedness or flatness around the mode of a frequency distribution. If a distribution has high kurtosis, then it would have a taller peak and fatter tails, whereas a distribution with low kurtosis would have a lower and rounded peak along with thinner or shorter tails. The normal distribution is called mesokurtic because it does not have any excess kurtosis (meso means "middle").

Figure 1.3 An example of a normal distribution

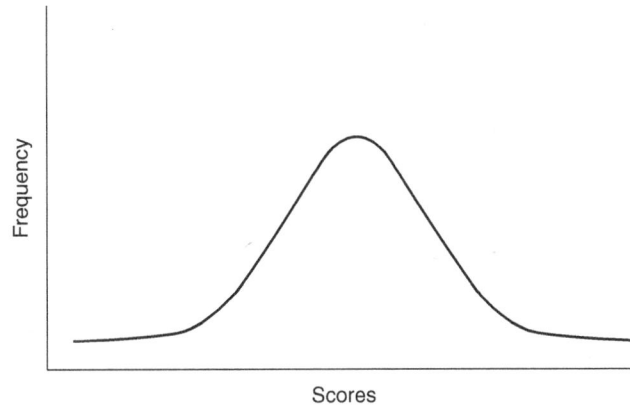

A normal distribution is often found in cognitive tests such as IQ, GRE, and the SAT. Moreover, it could be found in men's or women's height, blood pressure, weight, monthly rainfall totals over the course of years, and sometimes even students' grades (if we think of "C" as a mean). In psychology, we strive to obtain a normal distribution, but sometimes that just doesn't happen. Hence, there are other types of distributions that also occur depending on the situation. Some of these are also symmetric and have different types of kurtosis. For example, there is a leptokurtic (or leptokurtosis) distribution. A **leptokurtic** distribution, as shown in Figure 1.4, is one with more scores in the tails and fewer scores in the middle as compared to the corresponding normal distribution. In accordance with the definition of kurtosis, this distribution would have a high amount of kurtosis. Moreover, it tends to have a lower standard deviation. Often times, a leptokurtic distribution can be found with smaller sample sizes. The t-distribution (which we will address a little later) is leptokurtic. In a more practical example, the distribution of stock returns is also leptokurtic.

Figure 1.4 An example of a leptokurtic distribution

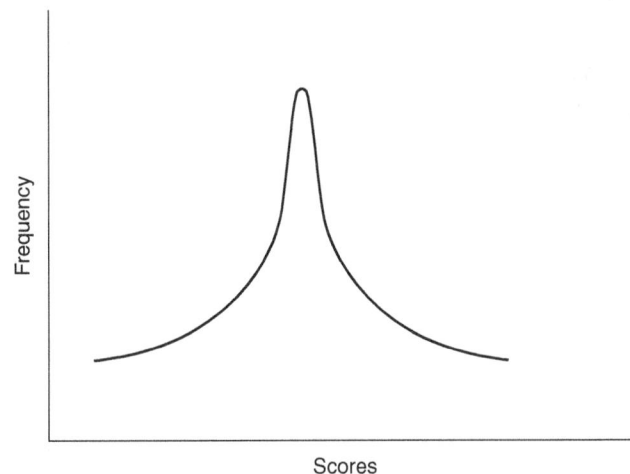

On the other hand, a **platykurtic** distribution, as shown in Figure 1.5, has fewer scores in the tails and more scores in the middle as compared to the corresponding normal distribution. As you can see below, this distribution is more flat and looks like the "duckbilled platypus." Naturally, this distribution would have a lower amount of kurtosis and it tends to have a larger standard deviation. One potential example of a platykurtic distribution (which would include a uniform distribution) might be the probability of roulette, keno, or lottery numbers occurring. If the numbers are chosen at random, then it is expected that the probability of each number occurring over time would be the same. For example, if there are 36 roulette numbers, then given 3600 spins of the wheel, it would be expected that each number would occur 100 times. In fact, if you go to a casino, you will see a board that shows the last 10 (or so) roulette numbers that were obtained. This is fairly useless information, given that there is the same probability of occurrence for all 36 numbers for each independent spin.

Figure 1.5 An example of a platykurtic distribution

Sometimes, not all distributions are symmetric. These distributions are skewed. **Skew** refers to the amount of asymmetry of the distribution. If a distribution is **positively skewed**, then the vast majority of the data is on the left (or low side) and the tail is pointing to the right. This is illustrated in Figure 1.6. One of the best examples of positive skew is the distribution of personal income. Of course, difficult tests are also an example of positive skew as most people score on the low end. However, if the measures of central tendency are around the peak for a symmetric distribution, then what happens with regard to an asymmetric one? Let us examine a difficult test. Suppose, you have five students who scored 10 out of 100 and one student who scored 100, then what is the mean? The mean would be 25. On the other hand, the median would be 10. Notice that the mean is skewed. This means that the one score (100) weighed the mean more to the right, so it would not be as good an overall indicator of central tendency as the median. In fact, when watching your local or national news, the announcer will say "median personal income," as statisticians knew many years ago that the best measure of central tendency for a skewed distribution was the median.

Figure 1.6 An example of a positively skewed distribution

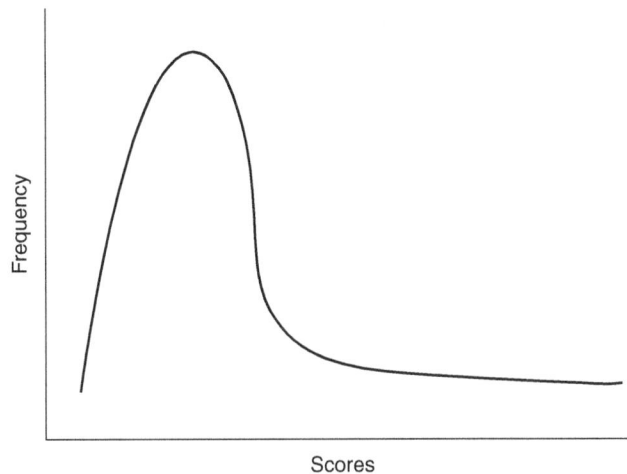

By comparison, if a distribution is **negatively skewed,** then the vast majority of the data is on the right (or high side) and the tail is pointing to the left. This is illustrated in Figure 1.7. In the following link, http://stats.stackexchange.com/questions/89179/real-life-examples-of-distributions-with-negative-skewness, examples of negative skew included the age of death of Australian males in 2012 and Men's Olympics long jump qualifying results in London from 2012. Moreover, age of retirement and easy tests (like what I give) are also good examples. Once again, you will find that the median will be the best measure of central tendency, as the mean will be skewed, given outliers. Hence, for skewed distributions, the three measures of central tendency are in different places, whereas for symmetric distributions, they are usually clustered around the peak.

Figure 1.7 An example of a negatively skewed distribution

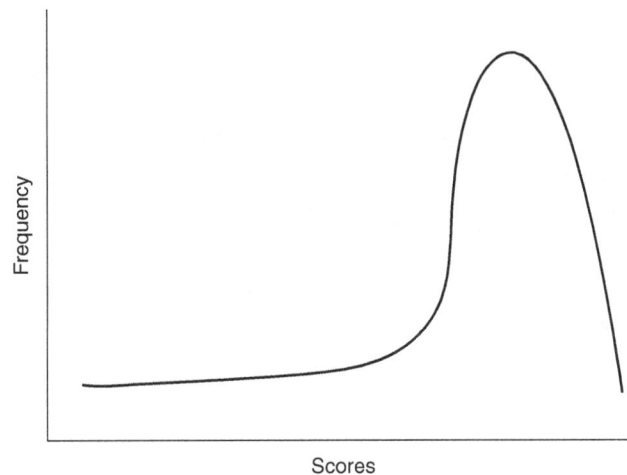

In general, these are the most common distributions found in psychology. However, I am asked on occasion about **bimodal distributions**. This distribution is illustrated in Figure 1.8. The only time that I have seen bimodal distributions is on a calculus test. There are people who really know what they are doing and others who are totally clueless. There are very few folks in the middle. In the next section, we will address what the sampling distribution of the mean looks like as we discuss the central limit theorem (CLT).

Figure 1.8 An example of a bimodal distribution

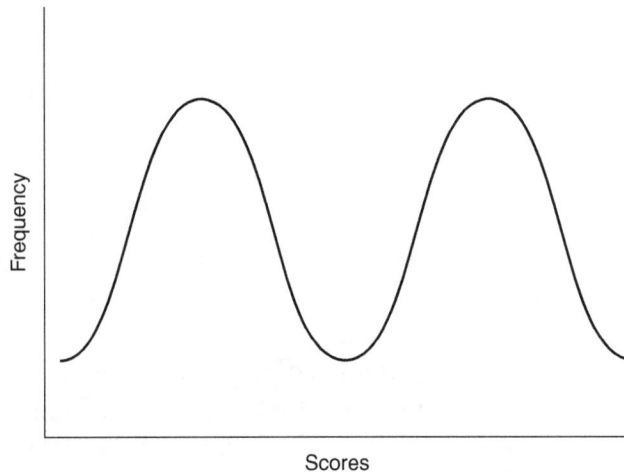

Central Limit Theorem

The CLT is a staple in the statistical field. Here it is in its entire splendor: "Given a population with finite mean μ and finite variance σ^2, the sampling distribution of the mean approaches a normal distribution with mean μ and variance σ^2/N, as N, the sample size, increases."

Let's break this apart to examine what it means. "Given a population with finite mean μ and finite variance σ^2", in mathematics, there is always something that is given. For example, even if one had one googol (which is one followed by 100 0's), then that would still be finite. Everything that we study in psychology (at least to my knowledge) has a finite mean and variance. The next part addresses the sampling distribution of the mean. First, a **sampling distribution** is a distribution under repeated sampling and equal-sized samples of any statistic. This can be shown empirically. For example, if you play keno, then you'll notice that there are 80 balls (ranging from 1 to 80). Suppose we randomly sample 10 keno balls and compute the mean. We throw them back into the hopper and randomly sample 10 keno balls again and compute the mean. If we do this about 5000 times, then we can create a frequency distribution that will look somewhat leptokurtic, as illustrated in Figure 1.9.

Figure 1.9 Sampling distribution of the mean of 10 keno balls over 5000 iterations

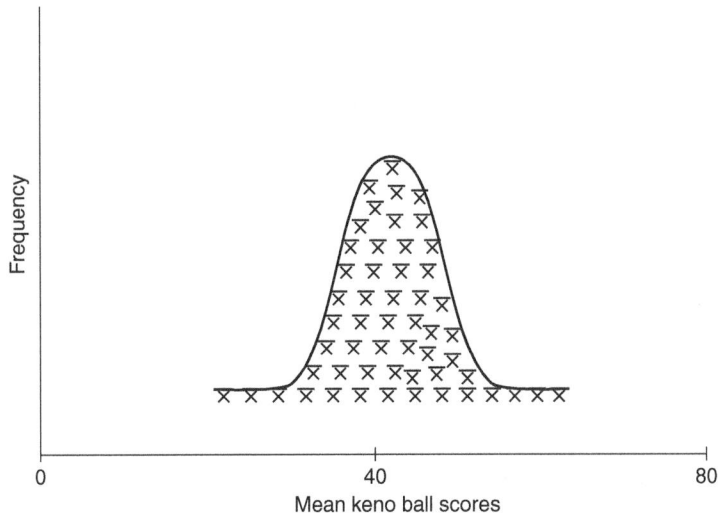

However, suppose we sample 30 keno balls and compute the mean. We throw them back into the hopper and randomly sample 30 keno balls again and compute the mean. If we do this 5000 times, then our frequency distribution will look more normal than the one which sampled only 10 keno balls, as illustrated in Figure 1.10.

Figure 1.10 Sampling distribution of the mean of 30 keno balls over 5000 iterations

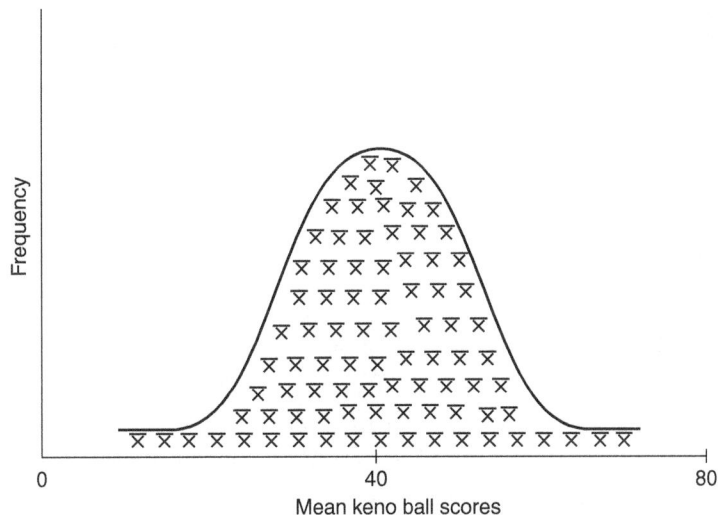

This would be the sampling distribution of the mean. The sampling distribution of the mean has a mean μ (the mean of the means if you know what I mean) and variance σ^2/N.

z-scores

Speaking of means, there is a statistic that allows us to determine how far we are away from the mean. That statistic is called a z-score. Hence, z-scores, or standard scores, determine how far away scores are from the mean in standard deviational units. A z-score may be both descriptive and inferential. For now, we'll consider it as a descriptive statistic. Before we address the formula for z-scores and how to do them, let's consider the following application. I'm sure that many of you want to go to graduate school albeit in psychology, hotel administration, biology, anthropology, sociology, nursing, or nutrition, just to name a few disciplines. In many areas, one needs to take the GRE (both verbal and quantitative), obtain three or more letters of recommendation, GPA, and provide a letter of intent. Suppose letters of recommendation and intent are rated from 1 to 5 (1 being horrible and 5 being excellent). Here are the hypothetical data for 10 potential applicants as shown in Table 1.1.

Table 1.1 Hypothetical data from potential graduate school applicants

	GRE Verbal (GREV)	GRE Quantitative (GREQ)	Letters of Recommendation (LOR)	GPA	Letters of Intent (LOI)
Applicant 1	148	151	4	3.2	3
Applicant 2	157	165	5	3.7	4
Applicant 3	164	164	3	3.3	4
Applicant 4	159	162	4	3.4	5
Applicant 5	156	158	2	3.1	1
Applicant 6	148	168	5	3.9	2
Applicant 7	152	171	4	3.8	5
Applicant 8	166	169	2	3.7	3
Applicant 9	162	153	3	3.5	4
Applicant 10	139	150	4	3.0	2

How should one evaluate these applicants? If we add everything and get a total score, then that will not work because the GREs are on a different scale than the other variables. In fact, if we did that, then they would be weighed about 30 times more than the other variables. That wouldn't make any sense if we believed that all variables should be weighed equally. Obviously, the correct answer has something to do with z-scores (which is what a number of universities do in evaluating candidates). Before I provide the answer, let's first examine what z-scores are.

The z-score is a transformational score. In short, regardless of the measurement, the z-score keeps scores in the same units (i.e., standard deviational units). If you are measuring height and weight, then we could provide a z-score for each person with regard to their height and weight. Without the z-score, we would be evaluating people on two different measurements. This is like comparing apples and oranges. Using z-scores makes the type of measurement irrelevant as z-scores keep everything in the same units.

Moreover, the distribution of z-scores is normal is normal. In fact, the distribution is called the standard normal distribution, which is depicted in Figure 1.11. It has a mean (or μ) of 0 and a standard deviation (or σ) of 1.

Figure 1.11 The standard normal distribution

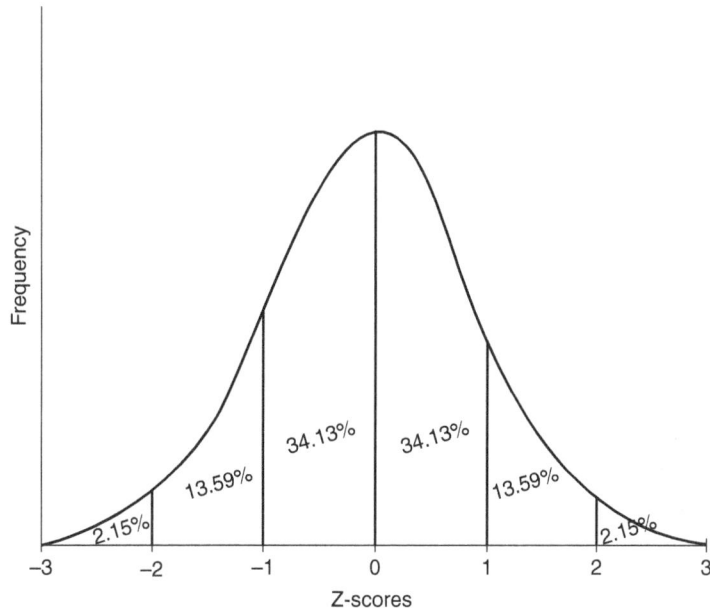

This does NOT mean that if you apply the z-score formula to a set of data that are skewed, you will make the skewed data become normal. Simply put, the z-score is a mathematical function in which if you put skewed data in, then you will get skewed data out. If you want to normalize the data, then you would need a nonlinear transformation such as a reciprocal, logarithm, square root, or otherwise. There are statistical software programs that can evaluate the data and perform specific transformations based upon the skewness and kurtosis. However, that discussion is beyond our course.

The formula is as follows: $z = (X - \mu)/\sigma$, in which X is the raw score, μ is the population mean, and σ is the population standard deviation. The z-score allows one to examine percentages and percentile ranks. Let's provide an example to see how. Suppose you took the Miller Analogies Test (MAT) for your graduate school (good news, the UNLV psychology department does not require it). In case you are interested, a question on the MAT might be Handel: Messiah :: Beethoven: (a) 5th symphony; (b) La Boheme; (c) New World Symphony; (d) Prelude to Die Meistersinger. Some of the questions delve into culture (a is the correct answer). The MAT, which ranges from 200 to 600, has a mean of 400 and a standard deviation of 25.

a) What percentage would an individual beat if they scored 410?

1. To obtain the z-score: $z = (410 - 400)/25 = +.40$

2. Draw a picture, notice that 0 represents the mean. This means that 50% of the individuals scored below 400 and 50% scored above 400.

Figure 1.11a What percentage would an individual beat if they scored 410?

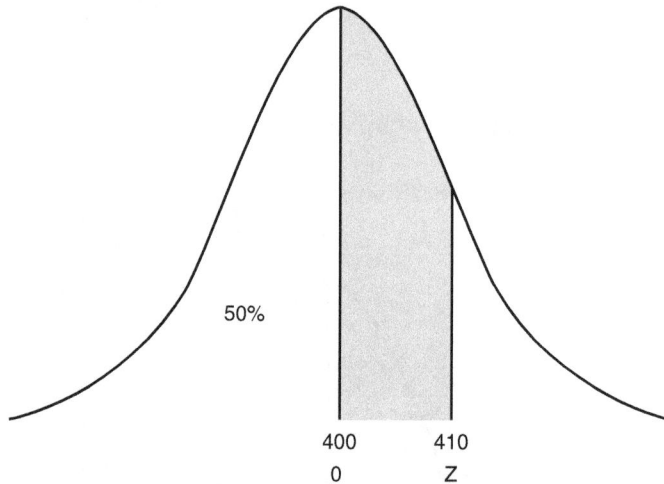

3. Go to the z-table. In the table, you will see three columns. The first column represents the z-score, the second column represents the percentage from the mean to z, and the third column represents the percentage from the z-score to the end of the tail. Notice that when you add the second and third columns together, you will get 50% or half the distribution. You will see that a z-score of .40 has .1554 between the mean and z and .3446 between the z-score and the tail.

4. We know that the individual scored higher than 50% of the people. Now we add .1554 or 15.54% which was between the mean and the z of .40. This gives us a percentage of 65.54%. Hence, the individual scored higher than 65.54% of the people.

b) What percentage would an individual beat if they scored 360?

1. To obtain the z-score: z = (360 − 400)/25 = −1.60

2. Draw a picture. Given the z-score, it is apparent that the individual beat less than 50% of the people.

Figure 1.11b What percentage would an individual beat if they scored 360?

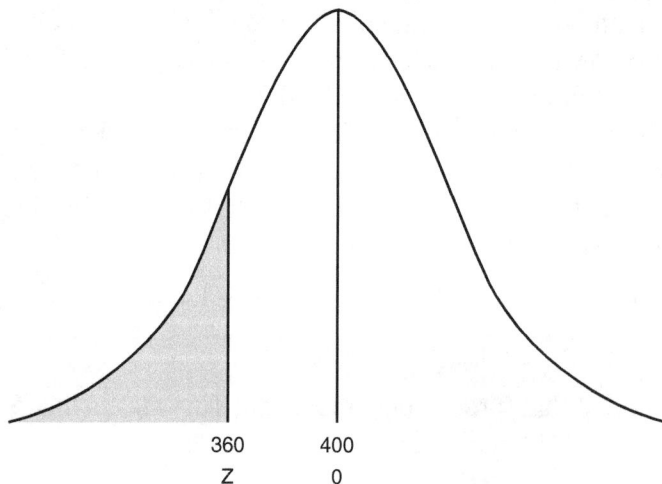

3. Go to the z-table. You will see that a z-score of -1.60 has .4452 between the mean and z and .0558 between the z-score and the tail.

4. Hence, the individual scored higher than 5.58% of the people. We could also say that 94.42% of the people who took the MAT beat them.

c) What percentage of the scores fall between 380 and 430?

1. This is a case in which you have to find the percentage of each z-score separately.

$$z = (380 - 400)/25 = -.80; z = (430 - 400)/25 = +1.20$$

2. Draw a picture.

Figure 1.11c What percentage of scores fall between 380 and 430?

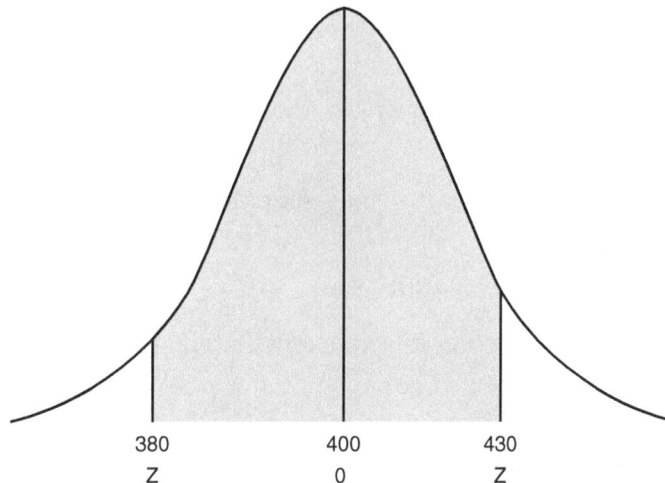

| 380 | 400 | 430 |
| Z | 0 | Z |

3. Go to the z-table. For the z-score of $-.80$, .2881 falls between the mean and z, whereas .2119 falls between the z and the tail. For the z-score of $+1.20$, .3849 falls between the mean and z, whereas .1151 falls between the z and the tail.

4. Based on the picture, add .2881 + .3849. Therefore, 67.30% of the scores should fall between 380 and 430.

d) Seven percent of the scores fall above what score?

1. If I reword the question, it is the same as saying 93% of the scores fall beneath what number? Instantaneously, you know that it is above 50%. We know the mean and standard deviation, but we know neither the score nor the z-score at this point.

2. You will notice that from the mean to z is .4300 and from the z to the tail is .0700. As we have two unknowns at this point $z = (x-400)/25$, the question becomes what z-score corresponds to the picture that we have drawn? The z-score is between $+1.47$ and $+1.48$, so we could say 1.475.

Figure 1.11d Seven percent of the scores fall above what score?

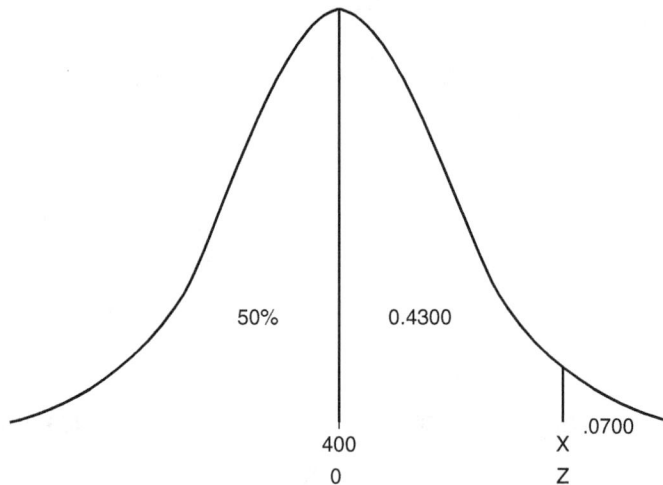

50% 0.4300

.0700

400   X
0   Z

3. $1.475 = (x - 400)/25$, now solve for x; $x = 436.875$. For all practical purposes, one could say 437.

e) How many individuals would one beat if they scored 405? There were 800 individuals who took the MAT.

1. To obtain the z-score: $z = (405 - 400)/25 = +.20$

2. Draw a picture. Given the z-score, it is apparent that the individual beat slightly more than 50% of the people.

Figure 1.11e How many individuals would one beat if they scored 405?

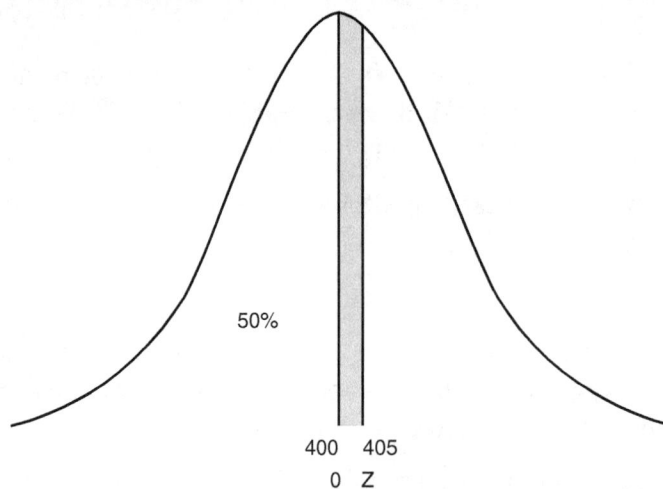

50%

400   405
0   Z

3. Go to the z-table. You will see that a z-score of +.20 has .0793 between the mean and z and .4207 between the z-score and the tail.

4. Hence, the individual scored higher than 50% + 7.93% or 57.93% of the people.

5. However, that wasn't the question. The question is how many individuals would they beat? To obtain this, multiply the percentage by the number of folks who took the MAT: $800 \times .5793 = 463.44$ people. For all practical purposes, one could say 463 people.

In our initial problem, we wanted to evaluate the applicants, given differing measurement types. As I mentioned, changing the raw data to z-scores allows you to keep the measurement units the same, and then you can weigh them as you see fit. In this case, I weighed them equally. That is, I took the average of the five z-scores as shown in Table 1.2. Just to refresh your memory, to obtain the z-score for the GRE Verbal for applicant 1, take the raw score (148) − the mean GRE V (155) and divide by the population standard deviation (7.96), which yields the z-score of −.89. If you add the z-scores for all applicants in the GREV category, then it will equal 0 (it might equal something slightly different because I only used two decimal places. If I carried out all the decimals, then I would obtain 0). To obtain the average z, I added all the z-scores and divided by 5 (that would weigh all z-scores equally or 20% for each). For applicant 1 ($-.89 + -1.38 + .39 + -.93 + -.23 = -3.04/5 = -.60$). One way to examine the data would be to eliminate everyone with an overall z-score below 0 (which is the mean). These would be applicants 1, 5, and 10. Those folks will receive rejection letters. If I had only four GA ships to offer, then the top choices would be applicants 7, 2, 3, and 4. Applicants 6, 8, and 9 might receive admission but no GA ship. One could also weigh z-scores differently. For example, if I wanted to weigh the GPA and LOI at 30% each, the GREV and GREQ at 15% each, and the LOR at 10% (it must all add up to 100%), then applicant 1's average z would be as follows ($-.89 \times .15$) + ($-1.38 \times .15$) + ($.39 \times .10$) + ($-.93 \times .30$) + ($-.23 \times .30$) = $-.64$.

The other question that is often asked of me is why I grade in z-scores. In my humble estimation, z-scores are the fairest way to grade. You will note that z-scores compare everyone to each other. It is what we call a norm-referenced standard. This is opposed to a criterion-referenced standard in which 90% is an "A," 80% is a "B," and so on. In the norm-referenced standard, it is irrelevant how easy or how difficult the quizzes are. Such would not be the case in the criterion-referenced standard. Obviously, if you have a difficult quiz, then the probability of obtaining 90% is problematic. As you can see, there are a plethora of uses for z-scores. On a fun and final note, I've always wanted athletes to have their contracts based on z-scores. For instance, a quarterback could state that their average z-score is +1.0 when examining total yards gained, percentage completion rate, average yards per pass attempt, and touchdowns.

Table 1.2 Hypothetical data from graduate school applicants including z-scores for each variable

	GREV	z-score	GREQ	z-score	LOR	z-score	GPA	z-score	LOI	z-score	Average z
Applicant 1	148	−.89	151	−1.38	4	+.39	3.2	−.93	3	−.23	−.60
Applicant 2	157	+.23	165	+.53	5	+1.37	3.7	+.62	4	+.55	+.66
Applicant 3	164	+1.11	164	+.39	3	−.58	3.3	+.62	4	+.55	+.41
Applicant 4	159	+.48	162	+.12	4	+.39	3.4	−.31	5	+1.33	+.40
Applicant 5	156	+.11	158	−.42	2	−1.56	3.1	−1.24	1	−1.81	−.98
Applicant 6	148	−.89	168	+.94	5	+1.37	3.9	+1.24	2	−1.02	+.32
Applicant 7	152	−.38	171	+1.35	4	+.39	3.8	+.93	5	+1.33	+.72
Applicant 8	166	+1.36	169	+1.08	2	−1.56	3.7	+.62	3	−.23	+.25
Applicant 9	162	+.86	153	−1.10	3	−.58	3.5	+1.24	4	+.55	+.19
Applicant 10	139	−2.02	150	−1.52	4	+.39	3.0	−1.55	2	−1.02	−1.15
Mean	155		161		3.6		3.5		3.3		
SD	7.96		7.3		1.019		.3225		1.2689		

z-score class problems

Given the SAT Verbal score in which the mean = 500 and the standard deviation = 100:

1. What percentage of folks did an individual beat by scoring 600?

2. What percentage of folks did an individual beat by scoring 370?

3. What percentage of folks fall between 420 and 560?

4. What score would 4% of the people be above?

5. What would be the total number of folks than an individual would beat if she scored 710? There were 200 people who took the exam.

Confidence Intervals

Suppose that I randomly sample 25 folks and administer a depression questionnaire (measured on a scale of 0–100, with 0 being no depression and 100 being extremely high depression). The mean of this sample was 48. Is this a good estimate of the population mean? Moreover, if someone did this study at a different university, then would their mean be in the same ballpark as my group? This is where confidence intervals come into play. A **confidence interval** is a 95%, 99% (or some stated) probability that the interval falls around (or about) the parameter. In laymen terms, we can think of a confidence interval as a large net that we are casting out in the ocean hoping to catch a fish. In this case, the fish would be analogous to the parameter (μ). So, we are trying to find the interval which will fall around μ. In order to do this, we need the following formula:

$$\overline{X} \pm t_{df}\, s/\sqrt{N}$$

$\overline{X}$ is the sample mean, s is the sample standard deviation, and N is the sample size. You will notice that we are dealing in sample statistics because we are trying to obtain our best estimate of μ. t_{df} represents the critical value of the t-distribution on a particular degrees of freedom (df). So, what are degrees of freedom? A **degree of freedom** is an independent piece of information. For example, suppose we have five numbers: 10, 20, 40, 60, 80. Moreover, suppose we are told that one of the numbers must equal 80 and the sum of the five numbers equals 210. Therefore, the other four numbers can vary any way they choose as long as the statement is satisfied. Hence, we could have 0, 0, 0, 130, 80; 20, 20, 60, 30, 80; etc. Four numbers are free to vary any way they see fit as long as the final number is 80 and the sum equals 210. Hence, there would be four (or N − 1) degrees of freedom. The formula for degrees of freedom is not always N − 1. For example, in correlation (which we will examine later), if we have 20 subjects and plot their height and weight, then we would want to get a line of best fit. As it takes two points to draw a straight line, the degrees of freedom in this case would be N − 2. That is, all points are allowed to vary along the line, provided that two points are fixed. Hence, degrees of freedom will change depending on the statistic. For confidence intervals, the degrees of freedom are N − 1. For our example, the sample mean is 48, the sample size is 25, the sample standard deviation is 5, and we are interested in the 95% confidence interval. To find the critical value of t, go to the t-distribution and at the top, we will look for a two-tailed test (which we will discuss later). Right under that is the probability. In this case, you will look for the .05 level (as 100% − 95% = 5%). Now, go down to 24 degrees of freedom (N − 1) and you will see the critical value being 2.064. So the equation is as follows:

$$48 \pm 2.064\left(5/\sqrt{25}\right)$$

Mathematically, $5/\sqrt{25} = 1$. So, we have 48 + 2.064 (1) and 48 − 2.064 (1). Therefore, the confidence interval written out is as follows:

$$\text{Prob } [45.936 < \mu < 50.064] \ .95.$$

This means that there is a 95% probability that this interval, from 45.936 to 50.064, will fall around (or about) μ.

If we want the 99% confidence interval, then everything would stay the same except for the t value. In order to find the t value, look at the level of significance for a two-tailed test being .01 (100% − 99% = 1%). Once again, go down 24 df and the critical value of t = 2.797.

$$48 \pm 2.797\left(5/\sqrt{25}\right)$$

Performing the computation similar to before, our answer would be

$$\text{Prob } [45.203 < \mu < 50.797] \ .99.$$

This means that there is a 99% probability that this interval, from 45.203 to 50.797, will fall around (or about) μ. When we increase from 95% to 99% probability, you will notice that the width of the interval gets larger 4.128 (50.064 − 45.936) versus 5.594 (50.797 − 45.203). Although this might sound counterintuitive, think of it in the following way: If someone asked you where Las Vegas is located and you weren't totally sure, you might be 95% positive that it is between Sloan and Apex and 99% positive that it is between Jean and Mesquite. The 99% positive might have a little more error to it (i.e., the difference between Sloan and Jean and between Apex and Mesquite) in order to accommodate the "uncertainty."

But what happens when you start manipulating sample size? How does that affect the width of the confidence interval? Let's suppose that we want a 95% confidence interval but increase our sample size from 25 to 64, keeping the mean at 48 and the standard deviation at 5.

$$48 \pm 2.064\left(5/\sqrt{64}\right)$$

Mathematically, $5/\sqrt{64} = .625$. Therefore, we have $48 + 2.064\,(.625)$ and $48 - 2.064\,(.625)$. Thus, the confidence interval written out is as follows:

$$\text{Prob}\,[46.71 < \mu < 49.29]\;.95.$$

This means that there is a 95% probability that this interval, from 46.71 to 49.29, will fall around (or about) μ. You will notice that if we increase our sample size, the width of the interval dropped from 4.128 (50.064 – 45.936) for our original interval to 2.58 (49.29 – 46.71). Hence, with greater sample size, the width drops, which means that we have greater accuracy. This should make total sense. Of course, as your N increases, you should get an estimate closer to μ.

What happens when you manipulate the standard deviation? How does that affect the width of the confidence interval? Let's suppose that we want a 95% confidence interval but increase our standard deviation from 5 to 10, keeping the mean at 48 and the sample size at 25.

$$48 \pm 2.064\left(10/\sqrt{25}\right)$$

Mathematically, $10/\sqrt{25} = 2.00$. So, we have $48 + 2.064\,(2)$ and $48 - 2.064\,(2)$. Therefore, the confidence interval written out is as follows:

$$\text{Prob}\,[43.872 < \mu < 52.128]\;.95.$$

This means that there is a 95% probability that this interval, from 43.872 to 52.128, will fall around (or about) μ. You will notice that if we increase our standard deviation, the width of the interval increases from 4.128 for our original interval to 8.256 (52.128 – 43.872). Hence, with greater standard deviation, the width increases, meaning that we have less accuracy. In this case, we increased our width by double, given that we doubled the standard deviation. Of course, if we have greater spread of the scores, then it only makes sense that accuracy would decrease.

Suppose that you wanted to reduce the interval to 1/2 of its original size, then what sample size would you need to do so? In order to address this question, we must obtain the interval first. Here is the interval from the initial problem.

$$2.064\left(5/\sqrt{25}\right) = 2.064$$

To reduce the interval to 1/2 of the original size, then here is the equation that we would need:

$$2.064\left(5/\sqrt{N}\right) = 1.032$$

Solving for N, N would equal 100. Hence, to reduce the interval to 1/2 of the original size, then we would multiply N by 4. Suppose you wanted to reduce the interval to 1/3 of the original size? Then what change in N would you need?

$$2.064\left(5/\sqrt{25}\right) = 2.064$$

Reducing the interval to 1/3, the following equation would be used:

$$2.064\left(5/\sqrt{N}\right) = .688$$

Solving for N, we would obtain 225. This would be nine times the original sample size. So, there is a trick here. To reduce the interval to 1/2 the size, take the denominator of the fraction (2) and square it for 4. That would be how many times you would need to multiply N by to obtain that interval. Likewise, to reduce the interval to 1/3 the size, take the denominator of the fraction (3) and square it for 9. That would be how many times you would need to multiply N by to obtain that interval.

One question that I am often asked is: If the degrees of freedom are between two numbers in the t-table, then what degrees of freedom should I use in order to determine the t value? Although standard statistical software packages compute confidence intervals precisely, which would make the question moot, I would offer the following recommendation: If you are between two degrees of freedom values, then choose the lower because you did not have enough sample size to obtain that extra bit of accuracy. Moreover, if $N \geq 100$, then use the critical value of t on infinity degrees of freedom (i.e., 1.96 or 2.58).

How does the central limit theorem relate to confidence intervals?

If we examine the formula, $\overline{X} \pm t_{df}\, s/\sqrt{N}$, there are really three pieces of the puzzle, namely, the sample mean, t_{df}, and $s/\sqrt{N}$. Let's take each piece and relate it back to the CLT.

How does the sample mean or $(\overline{X})$ relate? Going back to the CLT, the sampling distribution of the mean approaches a normal distribution with mean μ and variance σ^2/N, as the sample size increases. Notice that we are talking about the sampling distribution of the mean, not medians, not modes, so it only makes sense that we are dealing with the sample mean. Second, the sample mean is the best estimate of the population mean, μ.

The t-distribution is leptokurtic. However, as you add more sample size, the leptokurtic distribution becomes more normal. If the sample size were infinite, then if you look at your t-table (two-tailed test) on infinity degrees of freedom, you will find the critical values of 1.96 (at the .05 level) and 2.576 or 2.58 at the .01 level. Those are numbers that you might want to keep in long-term storage as we go throughout the semester. Do you remember what statistic is normally distributed? If you said the z-statistic, then you would be correct. If you examine the z-table, then what would be the z-score in which you would have 47.5% (.4750) between the mean and z and 2.5% (.0250) in the tail? Remember that the z-table encompasses only one side of the distribution. Of course, if it were two-tailed, then it would encompass 95% of the distribution (47.5% on each side) with 5% in the tails (2.5% in each). The z-score this would correspond to is 1.96. Likewise, if you examine the z-table with 49.5% (.4950) between the mean and z, and .5% (.0050) between the z and the tail (i.e., 99% encompassing both sides of the distribution and 1% encompassing both tails), then the z-score would fall between 2.57 and 2.58. Therefore, the relationship between t and z is as follows: $t^{\infty} = z$.

For the third piece of the puzzle, in accordance with the CLT, the variance of the sampling distribution of the mean is σ^2/N. If we square root the variance, then we obtain $\sigma/\sqrt{N}$. This is called the **standard error of the mean** for the population. A **standard error** is the standard deviation of a sampling distribution. Of course, when we square root the variance, we obtain a standard deviation. In this case, it is a very specific standard deviation; that is, the standard deviation of the sampling distribution of the mean (for the population). However, when we are computing confidence intervals, we don't have the parameter (μ). If we did, then we would not need a confidence interval. Instead, what we have is a sample trying to best estimate the parameter. Therefore, in the confidence interval formula (given that we have a sample mean), we need to have the standard error of the mean for the sample as well (in order to keep the equation congruent). Hence, that is why $s/\sqrt{N}$ is in the equation.

This is the end of our discussion of descriptive statistics. We will now move to the inferential statistics portion that addresses questions such as do the groups significantly differ on a particular measure?

Confidence Interval Class Problems

1. Suppose that you have a sample mean of 50, an N = 100, and a standard deviation of 10, what is the 95% confidence interval?

2. Suppose that you increase the N to 400, but leave the sample mean constant at 50 and the standard deviation at 10, what is the 95% confidence interval?

3. Suppose that you leave the sample mean at 50 and the N at 100, but you increase the standard deviation to 20, what is the 95% confidence interval?

Homework 1

The dreaded z-score and confidence interval lab

1. The following is a hypothetical set of scores from the Iowa Basic Skills Test by a 6th grade class. For the purpose of this lab, these scores will be considered a complete population, rather than a sample.

15	22	58	65	32	43	38	49
44	52	53	40	33	61	72	27

    a. Plot a frequency distribution of these scores (make sure that you are very specific in how you label your axes).

    b. Compute the median.

    c. Compute the mean.

    d. Compute the standard deviation.

    e. What percentage of scores should fall between the mean and 58, granted that the data are normal?

    f. How many scores should fall between the mean and 53?

    g. What percentage of scores should fall below 35?

    h. What percentage of scores should lie between 40 and 50?

    i. Two percent of the scores should lie above _____

2. Assuming that the above set of data were a sample from a population, find the 95% confidence interval on the mean.

3. What is the 99% confidence interval (using the same data)?

4. Assume that we cannot tolerate this interval width. If all else remains the same, what change in N would reduce this interval to 1/5 its size?

CHAPTER 2

Introduction to Statistical Significance Testing

SETTING THE SCENE

Inferential statistics are probably the most interesting facets of statistics, by far. It is here that we can make additional sense of the data in order to answer pertinent experimental questions. We begin by determining hypotheses. A **hypothesis** is an educated guess about which group will be significantly higher on a measure or if there will be a positive or negative relationship between two measurements. Here are some examples of hypotheses:

a) Children who are 10 years old will have significantly higher psychomotor vigilance task scores as compared to 7 year olds.

b) The elderly (aged 65 or older) will have significantly longer reaction times than adolescents (18 and younger).

c) Male college students will have significantly higher golf scores than female college students.

d) Attitudes toward police will be significantly lower in Ferguson, MO, than in Hoboken, NJ.

e) There will be a significant positive linear relationship between IQ scores and GPA.

In these cases, you will notice a prediction is made. It is not a good hypothesis to state that there will be a statistically significant difference between male and female college students on golf scores or that there will be a statistically significant difference between elderly and adolescents on reaction times. Moreover, you should not state that there will be a relationship between IQ and GPA. These latter statements beg the question: Which group will be higher or is the relationship a positive or negative one? In short, you have to put yourself on the line in terms of making predictions. Many times, we base our hypothesis on previous research, theory, experience, common sense, or even a gut hunch. Sometimes we can be wrong. But, do you know anyone who is omniscient? If I had a nickel for every time that I was incorrect in a hypothesis, then I would have retired years ago.

What are the components of the hypothesis? First, we start with an independent variable. An **independent variable** (IV) is a variable that is manipulated by the experimenter. For hypothesis (a), the independent variable is age, with two levels (10 and 7 year olds); for hypothesis (b), the independent variable is age with two levels (elderly and adolescent); for hypothesis (c), the independent variable is sex with two levels (male and female); and for hypothesis (d), the independent variable is city with two levels (Ferguson and Hoboken). Of course, you can have more than two levels of an independent variable (e.g., Ferguson, Hoboken, Las Vegas, and Reno). Here we would have four levels. We will address hypothesis (e) in more detail when we discuss correlation. For now, let's stay with the first four hypotheses. Along with an independent variable, the hypothesis also contains a dependent variable. A **dependent variable** (DV) is one that is measured or a score. For hypothesis (a), the dependent variable is the score on a psychomotor vigilance task; for hypothesis (b), the dependent variable is score on the reaction time task; for hypothesis (c), the dependent variable is golf scores; and for hypothesis (d), the dependent variable is scores on an attitude towards police measure. Here is the basic template for a hypothesis: Group A will score significantly higher than group B on the dependent variable.

Given the following hypotheses, could you determine what the independent variables and dependent variables are? Moreover, what are the levels of the independent variable in these examples? The Cincinnati Bengals will win more football games than the Pittsburgh Steelers in the AFC Central this year; UNLV students have significantly higher SAT scores than do students from UNR; brown rats will find their way to the goal box in an Olton Maze more often than will albino rats. Let's apply the concepts of a hypothesis to examining the types of errors that everyone makes when performing an experiment.

TYPE I AND TYPE II ERRORS

In any experiment, there are always inescapable errors. These errors exist regardless of how meticulous you are in research, what degree you have, how much experience you have in research, or even if you are a Nobel prize winner. Nevertheless, these are errors by which we all live with and try to keep under control. Let us examine the possible outcomes that can occur in research. For our example, we will delve into consumer psychology in which an experimenter was interested in determining if there were taste differences in chocolate chip cookies. Suppose we have three brands of cookies, namely, Keebler, Nabisco, and Pepperidge Farm. In this study, the participants looked at a paper plate with a cookie (no brand name, price, etc.), tasted it, and then rated it on a scale from 1 to 5 (with 1 being horrible tasting to 5 being absolutely wonderful). They brushed their teeth with a toothbrush (no toothpaste) and then sampled and rated a second cookie. Likewise, they repeated the same procedure for the third cookie. In accordance with research methods, we randomized the cookie order. If we did not and kept everything in the same order, then by the time you tasted the third cookie, your taste buds might be a bit tired. Thus, you might rank the third cookie lower in taste because you were tired of sampling. In this study, we start with a **null hypothesis** (designated as H_0). In a null hypothesis, the word null means no. Therefore, there is no statistically significant difference between or among population means on a particular measurement. In our example, the null hypothesis or H_0 would be there is no statistically significant difference among the population means of Keebler, Nabisco, and Pepperidge Farm chocolate chip cookies with regard to their taste ratings.

What would your hypothesis be with regard to chocolate chip cookie taste? One hypothesis might be that Pepperidge Farm would have a significantly higher taste rating than either Nabisco or Keebler. Moreover, Keebler would have a significantly higher taste rating than Nabisco. The theory might be one of price (if it costs more, then it is a better product). Of course, there are a number of such hypotheses, and your hypotheses would be equally as valid as mine. The proof is in the study.

Let's examine the study and determine the possible outcomes. First, suppose the null hypothesis is true, that is, the cookies are made out of the same ingredients (e.g., unbleached wheat flour, cocoa, butter, eggs), so in reality, they should all taste the same. However, when we conduct the study and perform the statistical analysis, we find that there are differences in taste ratings, when in reality, we should not find any. Is that OK or is that a mistake? Obviously, it is a mistake, or what we call Type I error. **Type I error** is the probability of rejecting a null hypothesis when in fact it is true. In essence, this is finding differences due to chance. Here is what I tell my graduate students about publishing: The more manuscripts you send out to journals, by Type I error alone, something is bound to get published. The more tickets you obtain in the lottery, then by Type I error alone, you are bound to get a winning ticket (i.e., $2). Type I error is indicated by the Greek letter alpha (α). In psychology, we tend to set alpha at .05. That means 5 times out of 100, the difference that we would obtain is by chance alone. Therefore, you can think of Type I error as a "false positive" as indicated in Table 2.1.

Suppose the null hypothesis is true; once again, the cookies are made out of the same ingredients (e.g., unbleached wheat flour, cocoa, butter, eggs), so in reality, they should all taste the same. When we conduct the study and perform the statistical analysis, we find that there are no statistically significant differences in taste ratings, when in reality, we should not find any. Is that OK or is that a mistake? It is OK or what is called $1 - \alpha$. $1 - \alpha$ is the probability of not rejecting the null hypothesis when it is true. The finding here would not be very exciting. It states that all the cookies taste the same because they are made out of the same ingredients. In this case, $1 - \alpha$ would equal .95 (if α equals .05). When you add the H_0 true cells together, they equal 1 or 100%.

There are some textbooks that will use the term accept H_0 rather than fail to reject H_0. You can never really accept a null hypothesis. For example, if my null hypothesis is he is not a dishonest man. I know that because the taxes are impeccable, clean, and totally honest. How do I know that he doesn't cheat on his wife? When you accept the null hypothesis, you believe that the null hypothesis is totally true. That would also be saying that the difference between population means would be 0. Of course, that hardly ever happens. By chance, there will be some difference. Hence, when you fail to reject the null hypothesis, there may indeed be no difference but there may also be a reasonable doubt or some other possibility that might also be feasible.

Suppose the null hypothesis is false. For example, there could be differences in the quality of ingredients. Maybe the cocoa from Pepperidge Farm cookies comes from Ghana, whereas the cocoa from Nabisco may come from Goodsprings, NV. Maybe Pepperidge Farm gets their wheat flour from Kansas, whereas Keebler may get theirs from Battle Mountain, NV. Hence, there should be a significant difference in taste ratings because of differences in quality (the null hypothesis is false), and after we perform the study and compute the analysis, we find that there was a statistically significant difference. Is that OK or is that a mistake? It is more than OK. This is what we call **power**. Power is the ability to detect a difference when, in fact, there is one. It could also be the ability of rejecting the null hypothesis when, in fact, it is false. It can also be thought of as the sensitivity of the experiment. So, we have at least three definitions for power. Let's take another quick example, so you can see how important power really is. For instance, if Pfizer creates a drug for eliminating Ebola, then you would hope that the symptomatology would significantly decrease compared to the control group (no drug) and ZMapp (another experimental drug). If that's what is found, then the chances are you have good power (and a nice payday).

Meanwhile, back to the chocolate chip cookies, suppose the null hypothesis is false and for whatever reason, when we perform our study, we do not find statistically significant differences, when in fact, we should find them. Then, we are dealing with **Type II error** or β (beta). This is the antithesis of power (which would be $1 - \beta$). Type II error is the inability to detect a difference when, in fact, there is one. It could also be a failure to reject the null hypothesis when in fact it is false. It can also be thought of as the insensitivity of the experiment. Again, we have three definitions for Type II error. In our experimental drug example, Type II error is that our drug should work, but for whatever reason, when we performed the statistical analysis, we did not find statistically significant differences in symptomatology among the three groups. We now focus on what are the determinants of alpha and power.

Table 2.1 The 2×2 table signifying the types of errors in an experiment

	Reject H₀	*Fail to Reject H₀*
H₀ true	α (Type I error) False positive	$1 - \alpha$ (true negative)
H₀ false	$1 - \beta$ (power) true positive	β (Type II error) false negative

Determinants of Alpha

Normally, in psychology, we set the overall Type I error rate to be .05. This is true of many academic disciplines. Specifically, it is you the experimenter and the journal editors who determine alpha. As you are aware, psychology has many journals (e.g., *Journal of Applied Psychology, Cognitive Psychology, Journal of Consulting and Clinical Psychology, Psychological Bulletin*) and it is those editors who set the alpha rates. Do not confuse journal editors with journalists. Journalists (i.e., reporters from the *Las Vegas Review Journal*, reporters on local TV stations) are probably not familiar with Type I error rates because they do not perform academic empirical research. Many of the scientific journal editors have conducted and published a multitude of studies in their field. In the "real world," which would include industry, advertising firms, and research organizations outside of academe, the determinants of alpha may be less stringent (as it is up to the experimenter). For example, there may be advertising firms claiming alpha rates of .20 to be statistically significant! Being an academic, that is too much chance for my taste. Obviously, the reasoning is to share with the client that their brand was significantly better than the other brands with regard to their dependent variable (which is what they wanted to hear and eventually publicize).

Determinants of Power

In this section, we examine how various factors influence power. Of course, some factors may be easier to control than others. Remember that power is the ability to detect a difference when, in fact, it exists. We would want to have power $(1 - \beta)$ around .80 (Howell, 2010). This indicates that we would find a difference that should be there (or rejecting the null hypothesis when it is false) approximately 80% of the time. Type II error (β) would be .20 or the percentage of time that we would not find the difference when the null hypothesis is false. One statistical test that we can use to determine if there is a difference between group means is the t-test (developed by William Sealy Gosset) . Gosset worked at Guinness brewery in order to determine the best yielding varieties of barley for beer. The formula for the t-test is as follows:

$$t = (\overline{X}_1 - \overline{X}_2) / (\sqrt{s_1^2 / n_1} + \sqrt{s_2^2 / n_2})$$

After calculating the t value (given means, variances, and sample sizes), you would compare it to the critical value in the t-table. If the calculated t value is higher than the critical value in the t-table, then it is statistically significant. This indicates that there would be a statistically significant difference between the means. If the calculated t value is smaller than the critical value in the t-table, then there is no statistically significant difference between the means. Hence, we would fail to reject the null hypothesis.

Given the formula for the t-test, let's examine some of the factors that influence power.

1. **Sample size**. As you can tell by the t-test formula, as the sample size increases, the denominator of the t-test will decrease, thereby increasing the value of the computed t. In case your mathematical skills leave a bit to be desired, let's take an example in which only the sample size is modified. Suppose the mean for group 1 is 9 and for group 2 is 5, the variance for group 1 is 3 and for group 2 is 6 and the sample size for both groups is 10.

$$t = (9 - 5) / (\sqrt{3/10} + \sqrt{6/10}) = 4.216$$

Suppose we keep everything constant except for sample size which we will increase from 10 to 100.

$$t = (9 - 5) / (\sqrt{3/100} + \sqrt{6/100}) = 13.33$$

As you can see, the computed t increases when we increase sample size. We can also demonstrate this from a different point of view. Let us examine the t-table. If you look at the .05 level (two-tailed), you will see the degrees of freedom on the left side. As you may recall, degrees of freedom are some form of sample size (in the confidence interval, it was N − 1). If you examine 15 degrees of freedom, the critical value is 2.131; whereas if you examine 20 degrees of freedom, the critical value is 2.086. Hence, it is easier to beat 2.086 than it is to beat 2.131. Smaller critical values (which are influenced by sample size) lead to greater power. Thus, as sample size increases, so does power, and Type II error decreases.

2. **Within Groups Variance** (σ^2). Because subjects are different within groups, there will be variability among those scores. Given that we are dealing with samples, the σ^2 would now become s^2. When we examine the t formula, as s^2 increases, the denominator would also increase, thereby leading to a lower t value. Once again, let's demonstrate that

$$t = (9 - 5)/(\sqrt{3/10} + \sqrt{6/10}) = 4.216$$

Suppose we increase the values of s^2 from 3 to 6 and from 6 to 9.

$$t = (9 - 5)/(\sqrt{6/10} + \sqrt{9/10}) = 3.265$$

As the within groups variance increases, then the t value drops. This, in turn, means that power drops and Type II error increases.

3. **Skew**. In general, the most power that one can obtain is when you have a normal distribution of data. Skew can be induced by increasing the variability of the data. For example, if you have the data 2, 4, 6, 8, 10, and if you square the data 4, 16, 36, 64, 100, you would increase the variability, which in turn, increases skew. Hence, as skew increases, the power drops and the Type II error would increase.

4. **Outliers.** An outlier is a piece of data that is well above or well beneath the mean of the distribution (e.g., + or −3 standard deviations from the mean). Examples of outliers would be Albert Einstein's IQ (which is about four standard deviations above the mean), LeBron James's income (about 600 standard deviations above the mean), or even the TV show Bonanza which ran 14 years (nowadays many shows are lucky if they last beyond one season). Of course, when you have outliers, you increase the within variability. Having outliers will decrease power and increase the Type II error. In fact, the last three determinants of power are fairly highly correlated.

5. **Difference between or among group means.** Let's suppose that we have five students with GPAs of 2.5, 2.6, 2.7, 2.8, and 2.9. The mean for this group is 2.7. Suppose that group 2 has a different five students with GPAs of 2.6, 2.7, 2.8, 2.9, and 3.0, for a mean of 2.8. As you can tell, there is little difference between these group means. However, suppose that group 1 had GPAs of 2.5, 2.6, 2.7, 2.8, and 2.9 for a mean of 2.7, whereas group 2 had GPAs of 3.9, 3.8, 3.9, 4.0, 3.9, for a mean of 3.9. It is conceivable that there would be a statistically significant difference in GPAs between these groups. Moreover, if you examine the numerator of the t-test, then you will see that the greater the difference between those means, then the higher the value of the computed t (given that the numerator will increase). Unlike the last three determinants, which deal with within groups variability, this deals with between or among groups variability. As the difference between or among group means increases, then power increases and Type II error drops.

6. **Alpha.** Remember that alpha (or Type I error) is the probability of rejecting the null hypothesis, when in fact, it is true. In order to see how it affects power, let's go back to the t-table. At the top

of the t-table, you will see probability for a two-tailed test at the .10, .05, and .01 levels. If we look at 20 degrees of freedom, then the critical values would be 1.725, 2.086, and 2.845, respectively. As the alpha level increases (e.g., .10), the critical value of t decreases, thereby making it easier to attain statistical significance (given that you want your calculated t value to be larger than the critical value of t). Hence, as alpha increases, power does as well, but somewhat "invalidly so" (Type II error would decrease). That is, in psychology and in many sciences, we set alpha to .05 (as I mentioned in the previous section). If we have a higher alpha, then that means we are taking advantage of chance occurrence. We may be obtaining "significant" differences, but they may be by chance alone.

7. **One-tailed versus Two-tailed tests**. Interestingly enough, this determinant concerns hypotheses, not necessarily distributions. In any experiment, there are three possible outcomes: (a) group 1 scored significantly higher than group 2 on the DV; (b) group 2 scored significantly higher than group 1 on the DV; and (c) there is no statistically significant difference between the population means of the two groups with regard to the DV.

Suppose a researcher in educational psychology was interested in examining the difference between SAT scores for incoming freshmen at CCSN and UNR. The researcher, who has conducted the same research for the past 10 years at these institutions, believes that UNR will have a significantly higher SAT score for the incoming class than does CCSN. In fact, the researcher is so indignant in this belief (as they found UNR has been significantly higher than CCSN in SAT scores over the past 10 years) that they believe the opposite (CCSN being significantly higher than UNR in SAT scores) could never occur. This is what we call a one-tailed test. In this test, you put all your eggs in one basket. If the antithesis does indeed occur, and you play the research game properly, you will need to throw out your data and start all over.

On the other hand, suppose the same researcher has the belief that UNR will have a significantly higher SAT score for the incoming class than does CCSN. However, the researcher, who knows they are not omniscient, believes that the opposite could also occur (i.e., CCSN being significantly higher than UNR in SAT scores). This would be thought of as a two-tailed test. In this case, you have a hypothesis, but hedge the bets a little realizing that the opposite might occur.

How does this affect power? Let's go back to the t-table to find out. Suppose you have 20 degrees of freedom, have a one-tailed test (at the .05 level), and are correct in your hypothesis. In this case, the critical value would be 1.725. If the computed t is in the hypothesized direction and it is higher than 1.725, then it is statistically significant. On the other hand, the two-tailed test (at the .05 level) has a critical value of 2.086. This is a little higher (and harder to beat) than the one-tailed test. Therefore, if you are correct in the one-tailed test, you gain all power. However, if you are incorrect in your hypothesis, then as you can see in Figure 2.1, the 95% contains both the null hypothesis and the alternative hypothesis (i.e., CCSN scores significantly higher than UNR on the SAT). So, you cannot see the forest through the trees, so to speak. All power is lost if you are incorrect in your hypothesis, so it is an all-or-none game for a one-tailed test.

Figure 2.1 An example of a one-tailed test

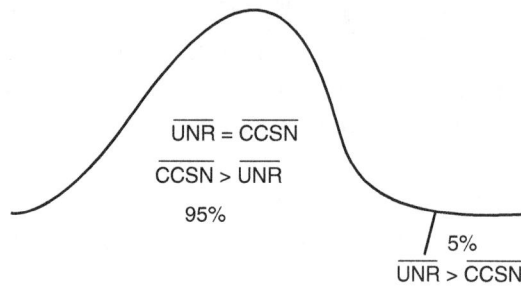

$$\overline{UNR} = \overline{CCSN}$$

$$\overline{CCSN} > \overline{UNR}$$

95%

5%

$$\overline{UNR} > \overline{CCSN}$$

In a two-tailed test, if you are correct, then you still have good power (critical value of 2.086 to beat rather than 1.725), albeit not as much power as if you are correct with a one-tailed test. However, if you are incorrect in your hypothesis, as you can see in Figure 2.2, then you are still able to see the alternative (i.e., incoming CCSN students have a significantly higher SAT score than UNR students). So, you have good power here (unlike the one-tailed test). By performing a two-tailed test, it is my humble opinion that you are covering your tail in case you are incorrect.

Figure 2.2 An example of a two-tailed test

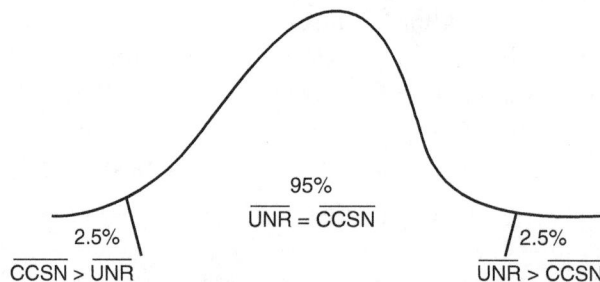

95%

$$\overline{UNR} = \overline{CCSN}$$

2.5%

$$\overline{CCSN} > \overline{UNR}$$

2.5%

$$\overline{UNR} > \overline{CCSN}$$

Today, most hypotheses are performed in two-tailed terms because none of us are omniscient. Besides, sometimes the most interesting findings are the ones that are antithetical to your beliefs. There are very few cases when a hypothesis could be one tailed. I suppose an example of a one-tailed hypothesis could be that the body will cure itself (i.e., no medication is necessary) if it is infected by Ebola. My humble belief (although I am certainly not an M.D.) is that one would die without proper treatment. Therefore, I cannot see survival as the alternative.

CHAPTER 3

One-Way
ANOVA—Two groups

ANALYSIS OF VARIANCE (ANOVA)—INTRODUCTION

As we begin our trek into inferential statistics, one of the questions that a researcher can postulate is: Is there a statistically significant difference between the means of group A and group B on some measure? We addressed this idea earlier when we set the scene concerning experimentation. Let's consider the following example.

A clinical psychologist was interested in examining the number of inappropriate affect responses between paranoid and disorganized schizophrenics at the Bellevue Hospital during a one-week period. The psychologist tested eight subjects in each group (e.g., S1 is Subject 1). The data are provided in Table 3.1.

Table 3.1 Hypothetical data for the number of inappropriate affect responses for paranoid and disorganized schizophrenics

	Paranoid		*Disorganized*
S1	9	S9	2
S2	11	S10	3
S3	8	S11	4
S4	14	S12	1
S5	3	S13	5
S6	7	S14	1
S7	10	S15	2
S8	8	S16	3
Totals	70		21
Mean	8.75		2.625

a) What is the independent variable in this example? Answer: Schizophrenia Type.

b) How many levels of the independent variable are there? Answer: 2.

c) What is the dependent variable in this example? Answer: Number of inappropriate affect responses.

d) What is the null hypothesis in words? Answer: There is no statistically significant difference between the population means of paranoid and disorganized schizophrenics with regard to their number of inappropriate affect responses.

e) What is the null hypothesis in symbols? Answer: $\mu_p = \mu_d$

Of course, we want to see if there is a statistically significant difference between these means. However, before we perform that task, it might be wise to examine Sir Ronald A. Fisher's rationale concerning analysis of variance or ANOVA, which is one way to examine whether this null hypothesis can be rejected or not.

THE RATIONALE FOR ANALYSIS OF VARIANCE: HOW DOES IT RELATE TO THE CENTRAL LIMIT THEOREM?

Here is our example which is a replication of Table 3.1:

	Paranoid		*Disorganized*
S1	9	S9	2
S2	11	S10	3
S3	8	S11	4
S4	14	S12	1
S5	3	S13	5
S6	7	S14	1
S7	10	S15	2
S8	8	S16	3
Totals	70		21
Mean	8.75		2.625

Using this problem, what was Fisher's thinking when he developed ANOVA?

Fisher knew that there was variability between the group means (in this case, between 8.75 and 2.625). One could certainly compute the variance between these groups. We define this as $s_{\bar{x}}^2$ or the variance among group means. Moreover, there is also variability within groups. After all, people differ. For example, looking at the paranoid group, there is variability among the numbers

9, 11, 8, 14, 3, 7, 10, and 8. One could compute a variance within that group. Furthermore, there is also variability in the data within the eight subjects of the disorganized group. Again, we could compute a variance here as well.

He also knew about CLT, and in case you forgot, here it is.

Given a population with finite mean μ and finite variance σ^2, the sampling distribution of the mean approaches a normal distribution with mean μ and variance σ^2/N, as N, the sample size, increases.

The question that Fisher asked is: Given that we are examining the differences between means and keeping in mind that we are dealing with samples trying to generalize to the population, then based on the CLT, what is the best estimate of σ^2? Is it the variance within the paranoid or within the disorganized schizophrenics? Obviously, if both groups are equally valid, then the best estimate would be to take the average of those within variances in order to obtain the best guess of σ^2. In statistical terms, the best guess of σ^2 is $\Sigma s_i^2/g$. Σs_i^2 represents summing the variances within each group (the subscript i is a descriptor indicating the number of groups that you are summing; it has no actual mathematical value); g is the number of groups. Hence, $\Sigma s_i^2/g$ means to sum the within variances and divide the number of groups, thereby giving you the average variance within groups.

In accordance with the CLT, the sampling distribution of the mean has a variance of σ^2/N. Fisher's best estimate of σ^2 was $\Sigma s_i^2/g$. However, in order to provide the full variance, we must account for sample size based on the CLT, $(\Sigma s_i^2/g)/n$. The reason that it is n (the number of participants in a group) is because we are dealing with samples as compared to a population (which is the optimum).

We now have two sources of variability, one due to the variability between (or among) group means $s_{\bar{x}}^2$ and one within groups: $(\Sigma s_i^2/g)/n$, so how can one analyze (or compare) them? One could subtract them to determine how large one is compared to another. However, Fisher chose to use a ratio format. If you have 6/4, then you know how much larger 6 is compared to 4 (1.5 times). Using that logic, here is the following equation: $(s_{\bar{x}}^2)/((\Sigma s_i^2/g)/n)$. I put parentheses around the specific terms so that you wouldn't be totally confused with all the division signs. As I mentioned, $s_{\bar{x}}^2$ is the variance between or among group means. Hence, it is evaluated in mean terms $(\Sigma X/n)$. The average variance within groups is evaluated as raw data $\Sigma s_i^2/g$ (similar to ΣX) and divided by n, which also puts it into mean units. Therefore, these variances are comparable; that is, comparing mean units to mean units. However, suppose we want to compare raw data units to raw data units? In that case, take the numerator $s_{\bar{x}}^2$, which is in mean units $(\Sigma X/n)$ and multiply by n. In the denominator, we can move the n into the numerator to create those raw data units. Therefore, the heuristic formula of ANOVA is the following: $(ns_{\bar{x}}^2)/(\Sigma s_i^2/g)$. In English, this formula is n (the number of folks in a group) times the variance between (or among) group means divided by the average variance within groups. In fact, when Fisher first derived out the formula, he called this z. Of course, he called everything z. It wasn't until statistician George Snedecor proclaimed that this should be called F in honor of Fisher. So, the heuristic formula for $F = (ns_{\bar{x}}^2)/(\Sigma s_i^2/g)$. Now that we have an understanding of Fisher's rationale, we can put the heuristic formula into practice.

HEURISTIC FORMULA FOR ANOVA

Let us take what we learned from the rationale of ANOVA and apply it to our problem.

	Paranoid		*Disorganized*
S1	9	S9	2
S2	11	S10	3
S3	8	S11	4
S4	14	S12	1
S5	3	S13	5
S6	7	S14	1
S7	10	S15	2
S8	8	S16	3
Totals	70		21
Mean	8.75		2.625

As you remember from the last section, the heuristic formula of $F = (ns_{\bar{x}}^2)/(\Sigma s_i^2/g)$. Let's begin with the numerator and the formula for determining the variance among (or between) group means ($s_{\bar{x}}^2$). Always make sure that you keep all the decimal places in the calculator when computing these formulas. You can always make them two decimal places at the end (and don't round up – you are not entitled to any extra power).

$$s_{\bar{x}}^2 = \frac{\Sigma \bar{X}^2 - \dfrac{\left(\Sigma \bar{X}\right)^2}{g}}{g - 1}$$

Step 1 Square each mean and then sum them

$8.75^2 + 2.625^2 = 83.453$

Step 2 Add the means and then sum them

$8.75 + 2.625 = 11.375$

Step 3 Square that number: $11.375^2 = 129.39$

Step 4 Divide that number by g (the number of groups) $129.39/2 = 64.695$

Step 5 Subtract step 4 from step 1: $83.453 - 64.695 = 18.758$

Step 6 Divide the number from step 5 by g – 1 (the number of groups –1). In this case, g – 1 = 1, so, $18.758/1 = 18.758$

Step 7 Multiply that number by n (the number of subjects in a group) to obtain the numerator, so $8 \times 18.758 = 150.064$

The numerator is 150.064.

To obtain the denominator, we must obtain the variance within each group and then average them. Hopefully, you will recall the computational formula for sample variance:

$$s^2 = \frac{\sum X^2 - \frac{(\sum X)^2}{N}}{N-1}$$

To obtain the sample variance for the paranoid group, you are now dealing with raw data, not means.

Step 1   Square each number and add them: $9^2 + 11^2 \ldots 8^2 = 684$

Step 2   Add each number and square that total $(70)^2 = 4900$

Step 3   Divide that number by N (the number of subjects in a group) $4900/8 = 612.5$

Step 4   Subtract the number from step 3 from the number from step 1: $684 - 612.5 = 71.5$

Step 5   Divide that number by N − 1 (which would be 7): $71.5/7 = 10.214$

Perform the same five steps for the disorganized group and you will end up with a sample variance equal to 1.9821

Now, average the within variances: $10.214 + 1.9821 = 12.196/2 = 6.098$

The denominator of the F ratio is 6.098.

So, to obtain F, it is $150.064/6.098 = 24.60$.

We will discuss what that F ratio means a bit later. But for now, make sure that you are comfortable with the procedure for the heuristic formula of F.

COMPUTATIONAL FORMULA FOR ANOVA

Before we can examine the mathematics of the computational formula, it is important that we define a few common (albeit archaic) terms from the statistical vernacular. **Sum of squares** is the numerator of a variance. If you go back to the variance formula, then you can think of it as $\Sigma(X - \overline{X})^2$ or $\sum X^2 - \frac{(\sum X)^2}{N}$. Hence, it is the sum of the deviational scores about the mean. A **mean square** is an estimate of population variance. Therefore, there will be a denominator of a variance somewhere in this formula.

We will need to compute three separate sums of squares in order to begin our computations. We will need the sum of squares of the independent variable (schizophrenia type), the within, and the total. When you add up the sum of squares of schizophrenia type with the sum of squares within, then you will equal the total sum of squares. Incidentally, you will never obtain a negative sum of squares (unless you compute incorrectly or have the wrong numbers in the equation).

To obtain the sum of squares schizophrenia type, here is the formula:

$$\frac{T_p^2 + T_d^2}{n} - \frac{(GT)^2}{N}$$

T stands for total. We are obtaining the total of the paranoid group (p) and the total for the disorganized group (d). n stands for the number of folks in a group. GT is the grand total. This can be obtained by adding the totals from the paranoid and disorganized groups. N is the total number of pieces of data. In fact, one can obtain n by taking N and dividing by the number of levels of the independent variable. Incorporating numbers, here are the SS (sum of squares) for schizophrenia type.

$$\frac{70^2 + 21^2}{8} - \frac{(91)^2}{16}$$

Step 1 $70^2 = 4900$; $21^2 = 441$, adding these totals equals 5341

Step 2   Divide $5341/8 = 667.625$

Step 3 $91^2 = 8281$

Step 4   Divide $8281/16 = 517.5625$

Step 5   Subtract step 4 from step 2: $667.625 - 517.5625 = 150.0625$

So, SS schizophrenia type $= 150.0625$

Incidentally, $(GT)^2/N$ is called the correction factor (CF). The **correction factor** takes the raw data and converts it into deviational scores. Deviational scores are raw scores minus the mean. Remember that sum of squares are $\Sigma(X - \overline{X})^2$.

Next, we need to obtain the sum of squares within. Normally, I do this last, as one formula is the sum of squares total (SS total) – sum of squares IV (or schizophrenia type, in this case). However, here is another formula:

$$\Sigma X^2 - \frac{T_p^2 + T_d^2}{n}$$

To obtain ΣX^2, square each raw score and add them together (i.e., $9^2 + 11^2 + \ldots 3^2$), this gives us 753.

$$753 - \frac{70^2 + 21^2}{8} = 85.375$$

In order to obtain the sum of squares total (SS total), use the following formula:

$$\Sigma X^2 - \frac{(GT)^2}{N}$$

Putting in numbers where applicable (as we have seen the portions of this formula before), we would obtain

$$753 - \frac{91^2}{16} = 235.437$$

If we examine the equations for the SS schizophrenia type, the SS within, and the SS total, you will notice that ΣX^2 is on the left (or positive) side of the equation for SS within and SS total, $(GT)^2/N$

is on the right (or negative) side of the equation for SS schizophrenia type and SS total. $\dfrac{T_p^2 + T_d^2}{n}$ is on the left side for the SS schizophrenia type and on the right side for the SS within, so they cancel out. In other words, if we take the ΣX^2 from the left side of SS within and $(GT)^2/N$ from the right side of the SS schizophrenia type and added those portions together, then you would get the equation for SS total.

So, to recap, the SS schizophrenia type = 150.0625, the SS within = 85.375, and the SS total = 235.437. Now, we need to place this into a summary table (which you will sometimes find in journals) in order to obtain the F ratio.

SUMMARY TABLE

Now that we have our sum of squares, let's begin by filling out our summary table, as shown in Table 3.2, in order to obtain F. At the top of the summary table, you will see the source. The source indicates what sums of squares we calculated. In this case, we calculated the sum of squares of schizophrenia type, within, and total. So, these are our three sources. df stands for the degrees of freedom. The degrees of freedom will change depending on the source (and I have a hunch that it could be the denominator of our F ratio, as alluded to in the previous section). SS is our sum of squares, MS is our mean square (variance), F is the F ratio, and p is the probability of the result not occurring by chance alone.

Table 3.2 Summary table for examining the differences in inappropriate affect responses as a function of schizophrenia type

Source	df	SS	MS	F	p
Schizophrenia Type	1	150.063	150.063	24.60	p < .01
Within	14	85.375	6.098		
Total	15	235.437			

Let's go through how we filled out the table. We start with the first source, which is schizophrenia type (the name of our independent variable). Some books might use the terms between or among rather than the name of the independent variable. This is poor practice for publishing in journals. The problem is that the journal reader (like you or me) must remember what the actual independent variable was. What happens if you have more than one independent variable? If you indicate "between" twice (two levels of each independent variable), then the reader is clueless as to which independent variable is which. In many journals that use summary tables, the name of the independent variable is indicated in the source column rather than the generic between or among groups. The degrees of freedom are g − 1. g represents the number of groups or levels of the independent variable. Here we have two groups (paranoid and disorganized), so 2 − 1 = 1. After filling in the sum of squares, the MS is the SS/df. Therefore, 150.063/1 = 150.063. We will come back to the F in a minute.

The next source is the within and the degrees of freedom here can be obtained in three ways:

a) df total – df schizophrenia type

   The df total = N – 1; N is the total number of pieces of data which are 16, so 16 – 1 = 15

   If we take 15 (df total) – 1(df schizophrenia type) = 14.

b) N – g, so here we have 16 – 2 = 14.

c) g(n – 1) in which g is the number of groups and n stands for the number of subjects within a group, so

   2(8 – 1) = 14

   Once again, the MS within can be obtained by taking the SS within and dividing by its df, so 85.375/14 = 6.098

The final source for the one-way ANOVA is the total. As mentioned, the df = N – 1 and the sum of squares has already been calculated. What is interesting here is that the SS formula is the numerator of the variance that we computed back when we discussed measures of dispersion.

$$\Sigma X^2 - \frac{(GT)^2}{N} \text{ or we can change the GT to the following:}$$

$$\Sigma X^2 - \frac{(\Sigma X)^2}{N}$$

When you divide this by N – 1, then you would obtain an MS total (or variance for the total). However, the MS total is never placed in the summary table, as it has no bearing on determining the F ratio.

In order to obtain the F ratio, take the MS schizophrenia type/MS within. Using numbers, it would be 150.063/6.098 = 24.60. You will notice that we are taking the ratio of the two variances (between groups variance/within groups variance). The F ratio indicates how many times larger the between (or schizophrenia, in this case) groups variance is to the within groups variance (or error variance). There is only one F ratio because we are only testing one null hypothesis (there is no statistically significant difference between the population means of the two schizophrenia groups with regard to the number of inappropriate affect responses). Thus, you get an F ratio for each null hypothesis tested.

How do you test for statistical significance?

What value do we put in for p? In order to find out, we need to go into the F table. This table contains the critical value of F for the .05 level and a separate table for the .01 level. Here is the game that we will play all semester long:

Is the F ratio that you compute larger than the critical value in the table?

a) If yes, then you reject the null hypothesis at the .05 or possibly .01 level (i.e., $p < .05$; $p < .01$). This means that there is a statistically significant difference between the group means. Therefore, one could state that group A scored significantly higher than group B on the DV.

b) If no, then you fail to reject the null hypothesis at the .05 level (i.e., p > .05; sometimes journals will use my initials n.s. to stand for nonsignificant). This means that there is no statistically significant difference between groups A and B on the DV.

In order to determine if the F ratio that we compute is larger than the critical value in the table, we start by going to the F table at the .05 level. At the top, it states degrees of freedom for the numerator. This corresponds to the degrees of freedom for the numerator of the F ratio (schizophrenia type). Therefore, the degrees of freedom for the numerator is 1. Along the column, you will notice that the table refers to the degrees of freedom for the denominator, within, or error term. In our example, the degrees of freedom within are 14. The critical value at the .05 level is 4.60. If we perform the same task for the F table at the .01 level, then the critical value would be 8.86. The F value that we computed 24.60 is larger than both 4.60 and 8.86, the critical values at the .05 and .01 levels, respectively. Hence, we would conclude that the p < .01.

How do we know that? Well, you will notice that the critical value and p have an inverse relationship as shown in Table 3.3. That is, as the critical value of F goes up, the p value goes down.

Table 3.3 The critical values for 1 and 14 degrees of freedom and the accompanying p values

Critical Value	*p*
4.60	.05
8.86	.01
Our F value = 24.60	much less than .01

After determining that the result is statistically significant, we now take a look at the means. The mean for paranoid is 8.75, whereas the mean for the disorganized schizophrenics is 2.625. Therefore, the conclusion that we can draw (and write in the journals) would be the following: There is a statistically significant higher number of inappropriate affect responses for paranoid than that for disorganized schizophrenics. Only stating that there is a statistically significant difference between the two groups with regard to inappropriate affect responses says nothing about which group is higher. That would be an incomplete story.

Let's examine a couple of similar hypothetical examples. Suppose we computed an F ratio of 6.16, then what would be the p value and the conclusion? In this example, 6.16 falls between the critical values of 4.60 and 8.86, so p < .05. There is an actual probability for 6.16 (p = .0263), but for our purposes, the probability falls between .05 and .01; hence, p < .05. The conclusion is still the same: There is a statistically significant higher number of inappropriate affect responses for paranoid than that for disorganized schizophrenics. The difference is with the certainty to base this conclusion. In this case, it is better than a 95% probability that this result did not happen by chance as compared to better than a 99% probability in our initial problem.

Suppose we computed an F ratio equal to 2.15. You will notice that this F ratio does not beat either 4.60 or 8.86; therefore, the p > .05. The conclusion here would be that there is no statistically significant difference between the paranoid and disorganized schizophrenics with regard to the number of inappropriate affect responses.

Students have often asked me throughout the years, "Can't we just look at the means and determine if they are statistically significant?" "There is certainly a big difference here". If you believe this, then suppose we have the following mean difference (mean of group A = 7.0 and mean of group B = 6.0). Is that a statistically significant difference (some of you may say "yes," whereas others might say "no"). The problem is that we need to all play by the same standard. At least Fisher has provided some standard for us to all play by.

One-way ANOVA (two groups)—
Class Example

A between or independent groups design; that is, only one score per subject.

A random sample of seniors in psychology was asked to give their overall GPA. There were six males and six females who participated. Here are the hypothetical results:

Subject	Male	Female
1	3.20	3.74
2	3.40	2.91
3	3.38	3.26
4	2.60	3.87
5	2.48	3.52
6	2.75	3.69
Total	17.81	20.99
Mean	2.96833	3.49833

Grand Total = 17.81 + 20.99 = 38.8

1. What is the null hypothesis?

2. What are the independent and dependent variables?

   a. what are the levels of the independent variable?

3. How do you compute the computational and heuristic formulas for F?

4. Know how to complete the summary table appropriately.

   a. how do you obtain the degrees of freedom?

   b. how do you obtain the sum of squares?

   c. how do you obtain a mean square?

   d. how do you obtain the F ratio and the probability associated with it?

5. Draw the appropriate conclusions to this study.

Homewotk 2

One-way ANOVA (two groups)

A social psychologist was interested in determining if there was a difference in Erotic Love scores as a function of internal versus external locus of control. The higher the score, the greater the Erotic Love. Your job is to perform a one-way ANOVA for both the heuristic and computational formulas. The hypothetical data are shown below.

Internal	External
44	53
42	47
46	48
35	44
36	68
45	55
40	50

Assumptions of ANOVA

When Fisher derived out ANOVA, he made a couple of assumptions. Let's examine these assumptions and explain why he made them.

1. **Normality in the Population**. Fisher made this assumption because that is how he derived out the critical values of F. Moreover, if we go back to the CLT, keep in mind that the sampling distribution of the mean is normal as the sample size increases. Given that we are dealing with means in an ANOVA (i.e., testing the difference between or among means), then it only makes sense that normality would be assumed. One can test this assumption with procedures such as the Shapiro–Wilk or the Kolmogorov–Smirnov. The Kolmogorov-Smirnov is a bit conservative (and I will discuss what that term means when we address violations of the assumptions), whereas the Shapiro-Wilk is considered more powerful and the better test to use for examining normality (Ahad et al., 2011).

2. **Homogeneity of variance (homoscedasticity) in the population**. This means that there are no statistically significant differences between or among the population variances (i.e., basically they are equivalent) within groups. In symbols, this assumption can be written as the following if we had three groups: $\sigma_1^2 = \sigma_2^2 = \sigma_3^2$. Why did Fisher make this assumption? Let's go back to the rationale of ANOVA. In the latter portion of the explanation, you will remember (I hope) that Fisher was trying to obtain his best estimate of σ^2 from the within-sample variances. If all sample variances are equally valid, then the best estimate of σ^2 was the average of the within sample variances ($\Sigma s_i^2/g$). Hence, you are averaging apples, apples, and apples, if the sample within variances are the same. If not, then you would be averaging unlike commodities like apples, coconuts, and pears. Of course, if you average those commodities, then it would mean nothing. There are tests that examine the homogeneity of variance in the population assumption as well. One of the best tests is Levene's test for medians (Brown & Forsythe, 1974; Conover, Johnson, & Johnson, 1981). There are others such as Levene's test for means and Bartlett's test just to name a couple.

Violation of the Assumptions

Most of the standard undergraduate textbooks will tell you that if you violate these assumptions, then the F-test is still robust. **Robust** means that if you violate the assumption(s), then the overall Type I error rate does not change appreciably from the nominal Type I error rate. For example, if we set Type I error equal to alpha or say .05 and if the test is robust, then the Type I error rate is still hovering around .05. This is exactly what you want. In fact, the best tests are robust and have good power. For our undergraduate purposes, saying that the F-test is robust with regard to violation of the assumptions is fine. Nevertheless, there is a bit more to the tale of which you should have some rudimentary knowledge.

If one violates the normality assumption, then the test might be a bit conservative. **Conservative** means that if you violate the assumption, then the overall Type I error rate is lower than the nominal (or stated) level. Normally, we set our Type I error rate to be .05. However, if a test is conservative, then the overall Type I error (or alpha) rate would be lower (e.g., .03). That means you would not be getting the power that you deserve. So you might not be finding differences that you should be able to obtain.

If you violate the homogeneity of variance assumption in a between or independent groups design (as we just talked about), then it is no problem. The F-test is still robust (the undergraduate answer). It should be noted that if you violate it because of skew, then the F-test will be conservative. However, if you violate it in a repeated measures design (in which a subject provides more than one score) in conjunction with unequal measure to measure correlation, then the F-test might be a bit liberal. **Liberal** means that when you violate the assumption, the Type I error rate is higher than the nominal level. In other words, if we set our overall Type I error to be .05, then a liberal test might have an overall Type I error rate of .10, for example. This means that we are finding statistically significant differences that might not be "true." Similar to conservative tests, liberal tests are not good either. Indeed, liberal and conservative are opposites of each other (as they are in politics too).

Characteristics of F

There are a number of characteristics of F that are interesting and unique. Moreover, many statistics overlap with each other mathematically. One of the characteristics of F will demonstrate that.

1. The F distribution is positively skewed. One can demonstrate this empirically. For example, suppose we tell a computer to create two groups with 10 subjects in each group. We could tell the computer to randomly select 10 bingo balls per group and calculate an F-test between the groups. If we did this about 10,000 times, then this might be the graphical representation as shown in Figure 3.1.

Figure 3.1 Sampling distribution of F with a sample size of 20

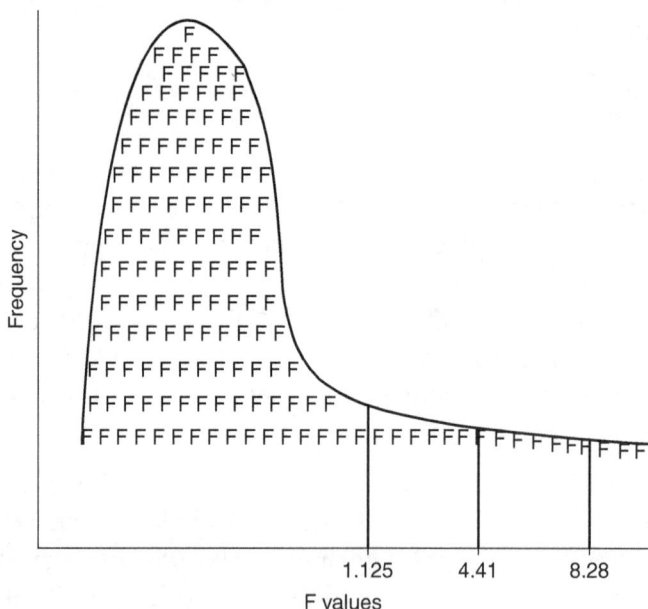

2. The mean of F = df within/df within − 2. In this example, our within degrees of freedom would be 18 (e.g., N − g; 20 − 2). Hence, if we produced 10,000 F values, then the mean F would be 18/16 = 1.125. If you examine the critical value of F at the .05 level on 1 and 18 degrees of freedom, you will see 4.41, whereas for .01 level, it would be 8.28. This means that 500 times out of 10,000 (5%), we would expect the F ratio to exceed 4.41 and 100 times out of 10,000 (1%), the F ratio should exceed 8.28. This means that many of the F ratios that we would generate would be less than 1.0.

3. Take a look at the F table with the critical value of infinity and infinity degrees of freedom. This means that you would have a population (an infinite number of groups coupled with an infinite sample size). You will see that the critical value is 1.0 for the .05 and the .01 levels. In fact, it is 1.0 for all probability levels (even .0000001). The importance of this is that if you obtain (or compute) an F ratio that is below 1.0, then there is no probability that the result will be statistically significant. This is a guarantee. Even my beloved Cincinnati Bengals winning the Super Bowl has a greater probability.

4. One of the more common tests for examining the difference between means is the t-test. The t-test formula is as follows:

$$t = (\overline{X}_p - \overline{X}_d) / (\sqrt{s_p^2 / n_p} + \sqrt{s_d^2 / n_d})$$

Using our example, $t = (8.75 - 2.625) / (\sqrt{10.214 / 8} + \sqrt{1.9821 / 8}) = 4.96$

If you take 4.96^2, you will find that it will equal our F ratio of 24.60. We can also demonstrate this in the t and F tables. If you examine the critical value for t on 14 degrees of freedom at the .05 level, then you will find 2.145. At the .01 level, the critical value of t would be 2.977. If you square each of those numbers, then you would obtain 4.60 and 8.86 respectively, which are the critical values of F on 1 and 14 degrees of freedom. Mathematically, $t_{df}^2 = F_{(1, df)}$.

From a graphic point of view, you will remember (I hope) that the t-distribution is leptokurtic (which is a symmetric distribution). If you square the t values that comprise the distribution, then you will add variability, which in turn will add skew to the distribution. Hence, the F distribution is positively skewed.

CHAPTER 4

One-Way ANOVA—more than two groups

We just went through the computations, summary table, and conclusions for a one-way ANOVA with two groups. However, there are myriad examples when a researcher wants to examine the difference among the population means of more than two groups. For instance, a social psychologist could address differences among Caucasians, African-Americans, Asians, and Hispanics on a financial risk scale. Moreover, a human resources specialist could examine the differences among football players, baseball players, basketball players, and hockey players on the Wonderlic Personnel Test. In this section, we begin by refreshing, reviewing, and reinforcing our knowledge with the heuristic and computational formulas along with determining the summary table and conclusion.

ONE-WAY ANOVA—MORE THAN TWO GROUPS—THE SCENARIO

A sensory psychologist was interested in examining the pleasantness of fragrance of three different perfumes: Chanel #5, Miss Dior, and American Beauty. The scale was rated from 1 (extremely unpleasant fragrance) to 5 (extremely pleasant fragrance). The results for the 21 subjects (S1 stands for Subject 1) are shown in Table 4.1:

Table 4.1 Hypothetical example of the pleasantness of fragrance of three different perfumes.

	Chanel #5		*Miss Dior*		*American Beauty*
S1	4	S8	3	S15	1
S2	5	S9	3	S16	1
S3	5	S10	3	S17	2
S4	4	S11	4	S18	2
S5	5	S12	2	S19	2
S6	5	S13	3	S20	1
S7	5	S14	1	S21	1
Totals	33		19		10
Mean	4.71		2.71		1.42

a) What is the independent variable in this example? Answer: Perfume type.

b) How many levels of the independent variable are there? Answer: Three levels.

c) What is the dependent variable in this example? Answer: Fragrance rating.

d) What is the null hypothesis in words? Answer: There is no statistically significant difference among the population means of Chanel #5, Miss Dior, and American Beauty perfumes with regard to their fragrance ratings.

e) What is the null hypothesis in symbols? Answer: $\mu_c = \mu_{md} = \mu_{ab}$

HEURISTIC FORMULA FOR ANOVA

The heuristic formula of $F = (ns_{\bar{x}}^2)/(\Sigma s_i^2/g)$. Once again, we will begin with the numerator and the formula for determining the variance among group means ($s_{\bar{x}}^2$).

$$s_{\bar{x}}^2 = \frac{\Sigma \bar{X}^2 - \dfrac{\left(\Sigma \bar{X}\right)^2}{g}}{g-1}$$

Step 1   Square each mean and then sum them

$4.71^2 + 2.71^2 + 1.42^2 = 31.632$

Step 2   Add the means and then sum them

$4.71 + 2.71 + 1.42 = 8.857$

Step 3   Square that number: $8.857^2 = 78.449$

Step 4   Divide that number by g (the number of groups): $78.449/3 = 26.1499$

Step 5   Subtract step 4 from step 1: $31.632 - 26.1499 = 5.482$

Step 6   Divide the number from step 5 by g – 1: $5.482/2 = 2.741$

Step 7   Multiply that number by n to obtain the numerator, so $7\,(2.741) = 19.19$.

The numerator is 19.19.

To obtain the denominator, once again, we must obtain the variance within each group and then average them.

The computational formula for sample variance is

$$s^2 = \frac{\Sigma X^2 - \dfrac{\left(\Sigma X\right)^2}{N}}{N-1}$$

To obtain the sample variance for the Chanel #5 group, you use the raw data, not the means.

Step 1   Square each number and add them: $4^2 + 5^2 + \ldots 5^2 = 157$

Step 2   Add each number and square that total: $(33)^2 = 1089$

Step 3   Divide that number by N (the number of subjects in a group): $1089/7 = 155.57142$

Step 4   Subtract the number from step 3 from the number from step 1: $157 - 155.57142 = 1.42858$

Step 5   Divide that number by $N - 1$ (which would be 6): $1.42858/6 = .2380966$

Perform the same five steps for the Miss Dior group and you will end up with a sample variance equal to .904762, whereas for the American Beauty group, the sample variance would be .2857143.

Now, average the within variances: $.2380966 + .904762 + .2857143 = 14.2857 / 3 = .476$.

The denominator of the F ratio is .476.

So to obtain F, it is $19.19/.476 = 40.31$.

Hopefully, this is a nice review for you. We move on to review the computational formula.

COMPUTATIONAL FORMULA FOR ANOVA

To obtain the sum of squares perfume type, here is the formula:

$$\frac{T_c^2 + T_{md}^2 + T_{ab}^2}{n} - \frac{(GT)^2}{N}$$

In reviewing the notation, T stands for total. This is the total of the Chanel #5 group (c), the total for the Miss Dior group (md) and the total for the American Beauty group (ab). n stands for the number of folks in a group. GT is the grand total. This can be obtained by adding the totals from the Chanel #5, Miss Dior, and American Beauty groups. N is the total number of pieces of data.

$$\frac{33^2 + 19^2 + 10^2}{7} - \frac{(62)^2}{21}$$

Step 1 $33^2 = 1089$, $19^2 = 361$, and $10^2 = 100$, adding these totals equals 1550

Step 2   Divide $1550/7 = 221.42857$

Step 3 $62^2 = 3844$

Step 4   Divide $3844/21 = 183.04761$

Step 5   Subtract step 4 from step 2: $221.42857 - 183.04761 = 38.38$

So, SS schizophrenia type $= 38.38$.

To obtain the sum of squares within, one can take the sum of squares total (SS total) – sum of squares IV (or perfume type). However, as we are computing it now, the formula is the following:

$$\sum X^2 - \frac{T_c^2 + T_{md}^2 + T_{ab}^2}{n}$$

To obtain $\sum X^2$, square each raw score and add them together (i.e., $5^2 + 5^2 + \ldots + 1^2$), this gives us 230.

$$230 - \frac{33^2 + 19^2 + 10^2}{7} = 8.571$$

In order to obtain the sum of squares total (SS total), use the following formula:

$$\sum X^2 - \frac{(GT)^2}{N}$$

Putting in numbers where applicable, we would obtain

$$230 - \frac{(62)^2}{21} = 46.95$$

So, to recap, the SS perfume type = 38.38, the SS within = 8.57, and the SS total = 46.952. Let's move to filling out the summary table.

THE SUMMARY TABLE

Let's review the pieces to the summary table. The source indicates what sums of squares we computed. In this case, we computed the sum of squares of perfume type, within, and total. So, these are our three sources. Df stands for the degrees of freedom. SS is our sum of squares, MS is our mean square (variance), F is the F ratio, and p is the probability of the result not occurring by chance alone.

Table 4.2 Summary table of fragrance ratings as a function of perfume type

Source	df	SS	MS	F	p
Perfume	2	38.38	19.19	40.31	p < .01
Within	18	8.57	.476		
Total	20	46.95			

We will review how we filled out the table. We start with the first source, which is Perfume or Perfume type. The degrees of freedom are g − 1. Here we have three groups (Chanel #5, Miss Dior, and American Beauty); therefore, 3 − 1 = 2. After filling in the sum of squares, the MS is the SS/df. Thus, 38.38/2 = 19.19.

The next source is the within, and the degrees of freedom here can be obtained in three ways:

1. df total – df perfume type

   The df total = N – 1; N is the total number of pieces of data which are 21, so 21 – 1 = 20

   If we take 20 (df total) – 2(df perfume) = 18.

2. N – g , so here we have 21 – 3 = 18.

3. g(n – 1) in which g is the number of groups and n stands for the number of subjects within a group, so 3(7 – 1) = 18

Once again, the MS within can be obtained by taking the SS within and dividing by its df, so 8.571/18 = .476.

In order to obtain the F ratio, take the MS perfume type/MS within. Using numbers, it would be 19.19/.476 = 40.31.

TESTING FOR STATISTICAL SIGNIFICANCE

Here is the review as to how to obtain the p value. First, go into the F table at the .05 and the .01 levels. Once again, here is the questions to ask:

Is the F ratio that you compute larger than the critical value in the table?

1. If yes, then you reject the null hypothesis at the .05 or possibly .01 level (i.e., p < .05; p < .01). This means that there is statistically significant difference between the group means. Therefore, one could state that group A scored significantly higher than group B on the DV.

2. If no, then you fail to reject the null hypothesis at the .05 level (i.e., p>.05; sometimes journals will use my initials n.s. to stand for nonsignificant). This means that there is no statistically significant difference between groups A and B on the DV.

In order to determine if the F ratio that we computed is larger than the critical value in the table, we start by going to the F table at the .05 level. At the top, it states degrees of freedom for the numerator. This corresponds to the degrees of freedom for the numerator of the F ratio (perfume type). Therefore, the degree of freedom for the numerator is 2. Along the column, you will notice the table refers to the degrees of freedom for the denominator, within, or error term. In our example, the degrees of freedom within are 18. The critical value at the .05 level is 3.55. If we perform the same task for the F table at the .01 level, then the critical value would be 6.01. The F value that we computed 40.31 is larger than both 3.55 and 6.01, the critical values at the .05 and .01 levels respectively. Hence, we would conclude that the p<.01.

After determining that the result is statistically significant, the only thing we can say at this point is that there is a statistically significant difference among the three perfumes with regard to their fragrance ratings. We might be able to say with some degree of definitiveness that Chanel #5 had a significantly higher fragrance rating than American Beauty. We can tell that on the basis of statistical significance and their means (Chanel # 5 = 4.71; American Beauty = 1.42). The problem is what about Miss Dior? Where does Miss Dior fall into the mix? If you were the CEO of Miss Dior, then the analysis as it stands doesn't tell you much. Therefore, we are getting an incomplete story by looking at just the omnibus (or overall) F-test. In order to get more information, we need to perform subsequent tests, so that we can get (as Paul Harvey would say), the rest of the story.

SUBSEQUENT TESTS TO ANOVA

Because we are only obtaining part of the story and it is somewhat nebulous at best (i.e., there is a statistically significant difference among the perfumes with regard to fragrance ratings), it is incumbent upon us to perform subsequent testing in order for us to determine which perfume fragrances are better than others. Hence, there will be three subsequent tests for us with which to start, namely, orthogonal comparisons, the Bonferroni test, and the Scheffe test.

Orthogonal Comparisons

Let's begin by providing the six characteristics of orthogonal comparisons and then we will put these into practice.

1. A priori hypotheses

2. $\Sigma a_i = 0$ (valid comparison)
   $\Sigma a_i b_i = 0$ (independence)

3. Each comparison is on 1 degree of freedom

4. You get as many comparisons as $g - 1$ degrees of freedom

5. Most powerful test of the three

6. Significance level = α

First, the hypotheses have to be done before any of the data are collected. Think of a research study as a blueprint to a house. The better you plan, the more stable the house. In this case, the more you plan ahead of time, the greater the power. Given a book or classroom, we will not be able to actually plan these comparisons in advance. However, in practice (making the assumption that people know about orthogonal comparisons), they would plan these hypotheses before collecting data.

In the second characteristic, the Σa_i means that when you add up the weightings, then they will equal 0. These weightings are obtained based upon your hypotheses. When you are performing orthogonal comparisons, the first hypothesis should include all the groups (no weightings should be 0). If you have a weighting of 0 for a particular group, then that means the group is not included in the hypothesis. My hypothesis is that the French perfumes (Chanel #5 and Miss Dior) will have significantly higher fragrance ratings than the American perfume (American Beauty).

Chanel #5	*Miss Dior*	*American Beauty*
−1	−1	+2

Notice that Chanel #5 and Miss Dior are weighted equally. It would make no sense to have one perfume weighed more than another, given that they comprise one group (i.e., French perfumes). American Beauty, however, is the other (separate) group. When you add up the coefficients, they will equal 0. It does not matter what coefficients are positive or negative. For example, one could also have the coefficients

Chanel #5	*Miss Dior*	*American Beauty*
+2	+2	−4

You could also have

Chanel #5	*Miss Dior*	*American Beauty*
1/32	1/32	−1/16

In all three examples, the hypothesis would be the same and you would eventually obtain the same sum of squares. Make sure when you combine groups that you give each group equal weighting. To determine the next comparison (given that we can only have two—see characteristic number 4), draw a line between the positive and negative coefficients (or weightings) as shown in Table 4.3.

Table 4.3 Orthogonal comparison of American Beauty versus (Chanel #5 and Miss Dior)

Chanel #5	*Miss Dior*		*American Beauty*
−1	−1			+2

The name of the game here is don't cross the line. Think of the line as a wall which cannot be crossed. As you can see based on Table 4.4, American Beauty cannot be compared with either Chanel #5 or Miss Dior because you would be crossing the line. American Beauty would have a weighting of 0 for the next comparison. Therefore, the only comparison that one can do would be Chanel #5 and Miss Dior.

Table 4.4 Set of orthogonal comparisons for the one-way ANOVA

Chanel #5	*Miss Dior*		*American Beauty*
−1	−1			+2
−1	+1			0

Once again, when you add up the coefficients (+1, −1, 0), they equal 0, so it is a valid comparison. Moreover, what has a positive weighting and what has a negative weighting is a moot point.

Of course, if I now draw a line between the positive and negative weightings, then there would be no other comparisons as I would be crossing the line(s). This is illustrated in Table 4.5.

Table 4.5 Set of orthogonal comparisons for the one-way ANOVA indicating the lines between positive and negative coefficients

Chanel #5	Miss Dior	American Beauty
−1	−1 |	+2
−1 |	+1 |	0

These comparisons not only need to be valid but must also exhibit independence as well. The word orthogonal means independent. You can think of independence as running separate and nonoverlapping studies. For instance, if person A runs a study at UNLV and person B runs the same study at UNR, then they are independent. However, if I provide my two-cents worth to both parties and they both use it, then these studies aren't totally independent, as they have my thinking involved. In a more layman scenario, if my wife and I are together, then we are not independent (as we are both dependent on each other for running a household). However, if she becomes my ex-wife and finds her happiness elsewhere and never sees me again (with no alimony involved), then we are now independent. In our example, we are looking for hypotheses that are independent. Independence means that there is no overlap in the hypotheses. The overlap here would be in the sum of squares. In other words, independent hypotheses will have sum of squares, that when added together, will equal the sum of squares of the independent variable. Before we demonstrate that, however, let's see if these comparisons are independent by definition ($\Sigma a_i b_i = 0$). The first hypothesis has the "a" coefficients, whereas the second hypothesis has the "b" coefficients, so we multiply the same column coefficients and then add.

Determining independence. To determine independence, multiply the coefficients from Chanel #5 (both the first and second hypotheses), add that to the product of the coefficients from Miss Dior and then add that to the product of the coefficients of American Beauty, so it looks like the following:

$$(-1)(-1) + (-1)(+1) + (2)(0) = 0, \text{ so indeed } \Sigma a_i b_i = 0.$$

We have shown that these two comparisons are not only valid but also independent. However, here is a small word of warning. Suppose you have more than two comparisons, then how do you determine independence?

It is not $\Sigma a_i b_i c_i d_i \ldots = 0$. Instead, you need to do all possible pairwise comparisons in which $\Sigma a_i b_i = 0$ for each.

Let's take the following scenario as shown in Table 4.6:

**Table 4.6 Orthogonal comparisons plus an extra
comparison for the one-way ANOVA**

Chanel #5	Miss Dior	American Beauty
−1	−1	+2
−1	+1	0
0	+1	−1

In order to determine independence, it is not $(-1)(-1)(0) + (-1)(1)(1) + (2)(0)(1)$, that is what is meant by $\Sigma a_i b_i c_i$. Instead, we would look at the first two hypotheses, which we know $\Sigma a_i b_i = 0$. However, if we look at the first and third hypotheses, then $(-1)(0) + -1(1) + (2)(-1) = -3$. Therefore, the entire set of comparisons is not independent or orthogonal. Of course, had this been equal to 0, then we would have examined the $\Sigma a_i b_i$ for the second and third hypotheses to make sure that it too was equal to 0.

To recap, here are the three steps for obtaining an orthogonal set of coefficients:

1. Make sure that you use all groups in the first hypothesis (i.e., no weighting of 0).

2. Draw a line between the positive and negative coefficients and do not cross the line.

3. Repeat Step 2 until you have drawn lines between all groups.

Now that we have established independence, we can move on to calculating the sum of squares for each hypothesis, obtaining the F ratios, and determining statistical significance. We must first calculate an intermediate step before computing the actual sum of squares.

$$L = \Sigma a_i T_i$$

L stands for a linear composite. This is because everything is multiplied to the first power exponentially. T is the total for each group. Here is how L is calculated for the first hypothesis:

Total = 33 Chanel #5	Total = 19 Miss Dior	Total = 10 American Beauty
−1	−1	+2
−1	+1	0

$$L = -1\,(33) + -1\,(19) + 2\,(10) = -32$$

To obtain the sum of squares, use the following formula:

$$SS = (L^2)/(n \, \Sigma a_i^2)$$

n is the number of subjects in a group and Σa_i^2 is the sum of the squared coefficients (similar to ΣX^2). To obtain Σa_i^2, square each coefficient and add them. Therefore, it becomes $-1^2 + -1^2 + 2^2 = 6$.

$$SS = (-32)^2/(7*6) = 24.38$$

For the second hypothesis, which is Chanel #5 v. Miss Dior, here is how to obtain the sum of squares.

$$L = -1(33) + 1 \, (19) + 0 \, (10) = -14$$

$$SS = (-14)^2/(7*2) = 14.0$$

Let's put this information into the summary table as shown in Table 4.7. The summary table is the same as our original one except that we are now adding our two subsequent hypotheses into the mix.

Table 4.7 Summary Table for Orthogonal Comparisons

Source	*df*	*SS*	*MS*	*F*	*p*
Perfume	2	38.38	19.19	40.31	p < .01
American versus French	1	24.38	24.38	51.21	p < .01
Chanel #5 versus Miss Dior	1	14.00	14.00	29.41	p < .01
Within	18	8.57	.476		
Total	20	46.95			

In our first hypothesis, the American perfume (American Beauty) versus the French perfume (Chanel #5 and Miss Dior) is on 1 degree of freedom (characteristic number 3). Again, obtaining the mean square is taking the sum of squares and dividing by the degrees of freedom, so 24.38/1 = 24.38. To obtain the F ratio, take the MS and divide by the MS within. Hence, 24.38/.476 = 51.21. Go into the F table on 1 and 18 degrees of freedom (critical values are 4.41 at the .05 and 8.28 at the .01 levels). Obviously, 51.21 is larger than both critical values, so the p < .01. In order to obtain the conclusion, we need to check the means. The mean of American Beauty is 1.42. Although you could average the Chanel # 5 and Miss Dior means (because the sample sizes are equal for each group), may I suggest the following:

If the mean is $\Sigma X/N$, then ΣX for Chanel #5 and Miss Dior would be 33 + 19 = 52. If we divide by N, then the mean is 52/14 = 3.71. Therefore, the conclusion would be there is a significantly higher fragrance rating for French perfume than for American perfume.

Likewise, if we examine the second hypothesis, we note that the MS is 14.00. Dividing this by the MS within gives us an F ratio of 14.00/.476 = 29.41. This F ratio is larger than the critical values (4.41 and 8.28, respectively), so p < .01. Again, examining the means (Chanel # 5 = 4.71; Miss Dior = 2.71), we can conclude that Chanel #5 had a significantly higher fragrance rating than did Miss Dior.

Aside from this esoteric $\Sigma a_i b_i = 0$, how else can we tell if these comparisons are independent? Notice if you add the degrees of freedom (1 + 1) it equals 2, or the number of degrees of freedom for perfume type. If you add the sum of squares for the comparisons (24.38 + 14.00), they will equal the sum of squares of perfume type. Therefore, what we have done is partition the perfume sum of squares into two independent portions, one for the American versus French perfume and the other being Chanel #5 versus Miss Dior.

Let me also point out that this is not the only orthogonal comparison possibility that you could have.

Table 4.8 A different set of orthogonal comparisons for perfume

Chanel #5	Miss Dior	American Beauty
+2	−1	−1
0	+1	−1

Table 4.8 illustrates another potential set of orthogonal comparisons. The difference is the hypothesis set. The hypotheses here would examine Chanel #5 versus the combination of Miss Dior and American Beauty and Miss Dior versus American Beauty. Statistically, this set of comparisons is as legitimate as our first set. Hence, it matters what you (the experimenter, your boss, or your faculty adviser) are interested in testing.

Because you are limiting the number of comparisons to g − 1, planning in advance, and adding the requirement of independence, you are gaining power. If you relax these restrictions, then you will lose power. However, there are times when you might need to do so. For example, suppose the CEO of Miss Dior wanted to know if they had a significantly higher fragrance rating than American Beauty, then the orthogonal coefficients that we initially chose would not fill the bill. In short, sometimes we might need to examine hypotheses that are not orthogonal. When that happens, then a feasible option would be the Bonferroni test.

Bonferroni

The Bonferroni test, named after the Italian mathematician Carlo Emilio Bonferroni, is another subsequent test option that has some relaxed restrictions. Here are the characteristics:

1. A priori hypotheses

2. $\Sigma a_i = 0$ (valid comparison)
   Don't need independence, so $\Sigma a_i b_i$ does not have to equal 0

3. Each comparison is on 1 degree of freedom

4. You get as many comparisons as you want, but you lose power with each subsequent comparison

5. Moderate power of the three

6. Significance level = α/c (c is the number of comparisons).

Similar to orthogonal comparisons, the hypotheses are done a priori, you still need a valid comparison, and each comparison is on 1 degree of freedom. However, unlike orthogonal comparisons, you do not need independence, you may make more than $g - 1$ comparisons, although you will pay in power for each subsequent one, and the significance level is now α/c .

In order to perform the Bonferroni test (and even the next test, the Scheffe test), the formulas for the sum of squares are the same as that of orthogonal comparisons. The only major difference between the Bonferroni test and orthogonal comparisons, as far as the summary table is concerned, is a potential change made to the significance level. You will notice that everything is the same through the F ratios. However, here is how the probability is determined. The significance level is α/c. In this case, c is 2, which represents the number of comparisons. For the .05 level, the significance level is .05/2 = .025. The critical value for 1 and 18 degrees of freedom at the .025 level is 5.97. For the .01 level, the significance level is .01/2 = .005. The critical value for 1 and 18 degrees of freedom at the .005 level is 10.21. Naturally, both F values (51.21 and 29.41) beat both critical values, so the ps < .01 (not .005). We need to use the original alpha level; that is, part of the loss of power. Note that the critical values for the Bonferroni test were 5.97 and 10.21 as compared to the orthogonal comparisons at 4.41 and 8.28. Given that it is harder to beat 5.97 and 10.21, we are losing a bit of power. Suppose you had four comparisons, then the critical values (.05/4 = .0125 and .01/4 = .0025) would be 7.69 and 12.32. As the number of comparisons increases, the critical values also increase, thereby making them harder to beat, leading to less power. In this case, the summary table (as shown in Table 4.9) is the same as the orthogonal comparisons and the conclusions are also the same. This is not always true, of course. Nevertheless, if you are going to perform the Bonferroni test, then make sure that you use a paucity of comparisons and choose wisely.

Table 4.9 Summary Table for the Bonferroni test

Source	*df*	*SS*	*MS*	*F*	*p*
Perfume	2	38.38	19.19	40.31	p < .01
American versus French	1	24.38	24.38	51.21	p < .01
Chanel #5 versus Miss Dior	1	14.00	14.00	29.41	p < .01
Within	18	8.57	.476		
Total	20	46.95			

Scheffe test

The American statistician, Henry Scheffe, developed a subsequent test that relaxes numerous restrictions, so that it is more versatile. However, one of the consequences of relaxing so many restrictions is that you pay in terms of power. Here are the characteristics:

1. A posteriori or post hoc hypotheses

2. $\Sigma a_i = 0$ (valid comparison)
   Don't need independence, so $\Sigma a_i b_i$ does not have to equal 0

3. Each comparison is on the among degrees of freedom

4. You can make as many comparisons as you want

5. Lowest power of the three tests

6. Significance level $= \alpha$.

Unlike the orthogonal comparisons and the Bonferroni test, the hypotheses may be done after the data are collected and each comparison is on the among degrees of freedom (as opposed to only 1). The latter is where you can really pay in power. Like the Bonferroni test, you do not need independence, but like all other subsequent tests, valid comparisons are a given. As mentioned previously, the sum of squares calculations are the same for all three subsequent tests. The major difference here is in the degrees of freedom for each comparison. As shown in Table 4.10, note that the degrees of freedom are the same as for perfume. Consequently, the mean square is divided by two for each comparison and the F ratios are half the size as the orthogonal or the Bonferroni comparisons. The degrees of freedom for each comparison are 2 and 18, so the critical values are 3.55 and 6.01 (the same degrees of freedom and critical values that we used for testing perfume type). Once again, both calculated F values are higher than the critical values, so ps < .01. You can imagine what happens if you had five different types of perfumes (4 degrees of freedom). For each comparison, the MS would be divided by 4, thereby making the F value one-quarter of the size of the orthogonal comparisons or the Bonferroni test. Therefore, this test might not be advantageous if you have a large number of groups.

Table 4.10 Summary Table for the Scheffe test

Source	df	SS	MS	F	p
Perfume	2	38.38	19.19	40.31	p < .01
American versus French	2	24.38	12.19	25.60	p < .01
Chanel #5 versus Miss Dior	2	14.00	7.00	14.70	p < .01
Within	18	8.57	.476		
Total	20	46.95			

One-way ANOVA
(three groups)—Class Example

An industrial psychologist was interested in determining if there are significant differences among three automobile companies with regard to the salaries of the mechanics. Here are the hypothetical results (these are the incomes presented in the thousands; that is, 43 really means $43,000). There were five mechanics samples from each company.

Subject	Toyota	Nissan	GM
1	43 *1849*	57	66
2	37 *1369*	52	72
3	28 *784*	54	75
4	32 *1024*	48	77
5	51	62	68
Total	191	273	358
Mean	38.2	54.6	71.6

Grand Total = 191 + 273 + 358 = 822

1. What is the null hypothesis?

2. What are the independent and dependent variables?

   a. What are the levels of the independent variable?

3. Know how to compute the ~~heuristic and~~ computational formulas for F?

4. Know how to obtain the summary table (e.g., sums of squares)

5. Know when and how to perform the tests subsequent to ANOVA (orthogonal comparisons, the Bonferroni test, and the Scheffe test)

Homework 3

One-way ANOVA (more than two groups)

A developmental psychologist was interested in examining the number of aggressive acts displayed by children after watching a certain type of movie. A child experienced one of the four situations:

1. No movie (control)

2. Viewed Snow White (only)

3. Viewed Alien (only)

4. Viewed Rambo (only)

   a. Your assignment is to perform a one-way ANOVA (show the summary table and all work). Draw the conclusions.

   b. You will also need to complete a set of orthogonal comparisons. The researcher was interested in testing whether seeing a movie would have more aggressive acts for children than not seeing one (control). Likewise, the researcher wanted to test whether the animated picture (Snow White) would yield less aggressive acts for children than the action films. There is also another comparison that will be left to you. Provide the summary table and conclusions.

   c. Use the same comparisons for the Scheffe test. Provide a separate summary table for the Scheffe test and draw conclusions from this study using this test.

Here are the hypothetical data:

Control	Snow White	Alien	Rambo
6	8	26	3
9	3	18	5
8	7	28	9
3	2	23	4
2	1	19	2
7	3	22	3
4	7	18	1
7	2	25	4
6	6	28	9
4	8	25	1

Experimentwise and Per Comparison Errors

Optimally, you want a test to have both a nominal Type I error rate and excellent power. There are a couple of different types of Type I errors. The first type is called **experimentwise error**. Experimentwise error is making at least one Type I error over the entire set of comparisons. Alternatively, experimentwise error is making at least one Type I error over the entire experiment. By contrast, **per comparison error** is making a Type I error for each individual comparison. If we examine our table of comparisons, then experimentwise error would be making at least one Type I error over the entire set (both comparisons in toto). It is an omnibus or overall Type I error rate (which we normally set at .05) for the overall experiment. The Type I error for the comparison of French versus American perfumes is a per comparison error (again, usually set at .05). Likewise, the Type I error for the comparison of Chanel #5 and Miss Dior is a per comparison error.

Chanel #5	*Miss Dior*	*American Beauty*
−1	−1	+2
−1	+1	0

How do the subsequent tests stack up in terms of experimentwise and per comparison error rates?

Table 4.11 Experimentwise and per comparison error rates for each of the subsequent tests

	Orthogonal Comparison	*Bonferroni's Test*	*Scheffe's Test*
Experimentwise error	$1 - (1 - \alpha)^{df}$	$\leq 1 - (1 - \alpha/c)^c$ $\leq \alpha$	α
Per comparison error	α	α/c	Quite small

Let's examine the per comparison error first. As you can see in Table 4.11, orthogonal comparisons have a per comparison error equal to α. This, of course, is optimal. As for the Bonferroni test, if you had only 1 comparison, then the per comparison error rate would equal α. However, if you had 3 comparisons, then the per comparison error rate drops to .0167 (assuming .05 /3). This would make the test a bit conservative. As for the Scheffe test, which is already conservative, the per comparison error rate is quite small and would probably fluctuate depending on the number of groups (hence, greater numbers of comparisons). Petrinovich and Hardyck (1967) demonstrated that given three groups and total sample sizes hovering between 15 and 90, the per comparison error rates for the Scheffe test were between .01 and .02, when they should have been .05. Interestingly enough, the ranking of the per comparison error rates mirror the rankings of power. That is, the most powerful being orthogonal comparisons, followed by the Bonferroni and the Scheffe tests. In per comparison error rate, orthogonal is optimal, the Bonferroni test is conservative, and the Scheffe test is even more conservative (in general). Of course, if you have a plethora of comparisons, then it is possible for Bonferroni's per comparison error rate to sink beneath that of the Scheffe test.

With regard to experimentwise error rate, you will notice that the Scheffe test is right on the money at α. Of course, the per comparison error is highly conservative, so we couldn't recommend the test. Suppose you had only 1 comparison for the Bonferroni test, then the experimentwise error rate would equal α. However, if you had 10 comparisons, then the experimentwise error rate would equal .0489. This is slightly less than α, but well within reasonable range of the nominal level (i.e., .05). For orthogonal comparisons, if you have only one comparison, then the experimentwise error rate is .05 (if $\alpha = .05$). However, if you have two comparisons, then the experimentwise error rate moves up to .0975. Although you might think that this is too liberal, it turns out that this is irrelevant, because the comparisons are independent. That is, it is similar to performing separate experiments.

The t-test revisited

Many textbooks spend numerous chapters discussing and computing t-tests. Although we have briefly introduced it more as a historical note and in its relation to F, you may wonder why we have discussed it in a perfunctory manner. When you have two groups, the t-test and the F-test are interchangeable. Both are perfectly fine to use. However, when you have three groups, the t-test becomes problematic. Take a look at the formula, once again.

$$t = (\overline{X}_1 - \overline{X}_2) / (\sqrt{s_1^2 / n_1} + \sqrt{s_2^2 / n_2})$$

As you can see, the formula compares only two means at a time. What happens when you have three means (i.e., the three types of perfumes)? Let's begin the discussion by remembering that the probability of making a Type I error is α. Obviously, $1 - \alpha$ is the probability of not making a Type I error. $1 - (1 - \alpha)$ would be the probability of making a Type I error again. Moreover, as shown in Table 4.12, let's assign coefficients to the three groups as if we were performing t-tests.

Table 4.12 All pairwise coefficients for doing all possible t-tests

Chanel #5	*Miss Dior*	*American Beauty*
+1	−1	0
+1	0	−1
0	+1	−1

If we are performing t-tests and can examine only two means at a time, then we would need to perform a t-test between Chanel #5 and Miss Dior, Chanel #5 and American Beauty, and Miss Dior and American Beauty. Hence, we would need three separate t-tests in order to examine the omnibus hypothesis. If the probability of not making a Type I error for each comparison is $1 - \alpha$, then the probability of not making a Type I error for all three t-tests would be $1 - \alpha^3$. This would make the Type I error equal to $1 - (1 - \alpha)^3$. If $\alpha = .05$, then the probability of making a Type I error for the three t-tests would be .1427. This is almost three times the amount of Type I error that we could tolerate. Of course, this is incredibly liberal, so you would never want to perform multiple t-tests because of this alpha inflation.

However, if we go back to the experimentwise error rate for orthogonal comparisons, it too was inflated. Yet, earlier I stated that it was irrelevant and no problem at all. For multiple t-tests, it is a problem. Why is there a potential double standard? If you examine the t-test comparisons, are they independent? Take the first two rows : $(+1) (+1) + (-1) (0) + (0) (-1) = +1$. These comparisons are not independent, so the entire set is not independent. That is what sets these two tests apart. Having an inflated Type I error rate for multiple t-tests is not tolerable because the comparisons are not independent. This is the reasoning as to why we don't talk about t-tests in great depth in this course. For two groups, t-tests are perfectly fine (and so is F), but beyond two groups, multiple t-tests have an inflated Type I error rate, whereas the F-test still has an experimentwise error rate hovering around alpha (as the test is fairly robust, at least that's my undergraduate answer). Thus far, we have talked about examining the levels of one independent variable in terms of statistical significance. But, what happens if we have two independent variables?

CHAPTER 5

Two-Way ANOVA—Between or Independent Groups

In a two-way or $A \times B$ design, we are dealing with two independent variables, rather than one. This creates more hypothesis testing. Of course, there will be a null hypothesis for each independent variable, but there is also an interaction between the two independent variables (which I will discuss a little later), which also has a null hypothesis. Therefore, in this design, there are three null hypotheses that we would need to test. The number of null hypotheses in a design (without subsequent tests) may be given by the formula $2^{IV} - 1$. Hence, when we had a one-way ANOVA, then we had $2^1 - 1 = 1$ null hypothesis. For this design, it would be $2^2 - 1 = 3$ null hypotheses. I know someone near and dear to my heart that had a five-way ANOVA for his Master's Thesis, so he had $2^5 - 1 = 31$ null hypotheses to test.

Let's begin by examining data from the following scenario: The political parties wanted to examine the differences in satisfaction of party across geographic regions. The satisfaction was rated on a scale from 1 to 10 with 1 being horrid and 10 being superlative. There were 28 subjects in total and each subject rated their party only once for their specific geographic region. For example, Subject 1, who is a Democrat from the east (east of the Mississippi River), rated their party an 8; whereas Subject 8, who was a Democrat from the west (west of the Mississippi River), rated it a 9. The data are shown in Table 5.1.

Table 5.1 Hypothetical data representing the satisfaction of the political parties as a function of party and geographic region

	Democrats		Republicans	
	East	*West*	*East*	*West*
	8	9	3	2
	5	8	4	8
	4	9	2	1
	7	10	4	1
	6	8	7	2
	4	9	1	1
	7	8	3	3
Totals	41	61	24	18
Means	5.857	8.714	3.428	2.571

This is a 2 × 2 independent groups design. There are two levels of the independent variable called political party (Democrats and Republicans) and two levels of geographic region (east and west). The dependent variable is the satisfaction rating.

NULL HYPOTHESES

As I mentioned, there are three in this design.

1. $\mu_D = \mu_R$

   There is no statistically significant difference between the population means of the two political parties with regard to their satisfaction rating.

2. $\mu_{east} = \mu_{west}$

   There is no statistically significant difference between the population means of the two geographic regions with regard to their satisfaction rating.

3. There is no political party × geographic region interaction in the population. (You would actually state verbally, that there is no political party by geographic region interaction in the population). Hence, × (or times) is substituted for the word "by" when verbalized.

Although there are symbols for the interaction hypothesis, it can get a little complicated, so I will not discuss them here. However, what is more important is the concept of interaction, so let's take a look.

Statistics primarily came from the agriculture area. As I mentioned previously, Gosset was interested in determining the best yielding malt barleys for the Guinness brewery. Suppose we have the following scenario in Table 5.2:

Table 5.2 Hypothetical scenario illustrating the yield of grain across different fertilizers

	F1	*F2*	
2-row Barley	70	0	70
6-row Barley	0	70	70
	70	70	

Here we have two different types of barleys (2-row and 6-row) along with two different types of fertilizers (F1 and F2) yielding a particular amount of grain (70 bushels). To an ordinary brewer, if you look at 2-row versus 6-row barley, then there is no difference in grain yield. Likewise, if you look down the columns of the fertilizers, then you would believe that there is no difference in grain yield. Hence, the brewer would conclude that there is no difference regardless of the type of barley planted and the type of fertilizer used. But, is that really the case? Obviously, the yield of the type of barley depends on the type of fertilizer used. For example, for 2-row barley, use fertilizer 1, whereas for 6-row barley, use fertilizer 2 for maximum yield. This constitutes an interaction. In a 2×2 design, like what we show here, if you cross multiply, you will notice a big discrepancy (4900 to 0), this will indicate an interaction. We can define an **interaction** as follows: The effect of an independent variable differs depending on the levels of the other independent variable. In our context, the yield of barley differs depending upon what type of fertilizer you use. In fact, obtaining a statistically significant interaction is one of the more interesting findings in science. However, in general, it is harder to find a statistically significant interaction than it does to find a significant difference among the levels of an independent variable (e.g., political party).

SUMS OF SQUARES

We will perform the sum of squares for each independent variable, the interaction, within, and total. The formulas are similar to the one-way ANOVA, but we are adding another independent variable and an interaction to the mix. Fortunately, we will not do this via the heuristic formula. If you are interested in performing a two-way ANOVA via means and standard deviations (similar to what we did with a one-way ANOVA), then check out Huck and Malgady's (1978) article.

$$SS \text{ political party} = (T^2_D + T^2_R)/n - (GT)^2/N$$

In order to obtain the totals, you must add over the geographic region. For Democrats, the totals would be 41 (from the east) + 61 (from the west) for a total of 102. Likewise, for Republicans, the totals would be 24 (from the east) + 18 (from the west) for a total of 42. The n is the number of pieces of data needed to obtain the 102 or 42. In this case, n = 14. There were 14 individual pieces of data making up 102 and 14 individual pieces of data making up 42. The grand total (GT) is 102 + 42 or 144. N is the total number of pieces of data, which is 28. Again, it could also be obtained by taking the number of the levels of the IV (2) times the number of pieces of data making up the 102 or 42, namely, 14. Therefore, $2 \times 14 = 28$. To refresh your memory, you could also take N/the number of levels of the IV in order to obtain n (28/2 = 14). This little trick always works as long as you have equal sample sizes within groups.

$$SS \text{ political party} = (102^2 + 42^2)/14 - (144)^2/28 = 128.571$$

We perform the same task for obtaining the SS geographic region.

$$SS \text{ geographic region} = (T^2_{east} + T^2_{west})/n - (GT)^2/N.$$

We obtain the totals here by adding over political party. For example, for the east, we take the total from Democrats (41) and add it to the total from Republicans (24), for a total of 65. Likewise, we perform the same task for the west: total of Democrats = 61 + total of Republicans = 18, for a total of 79. The n is still 14 (as there are 14 pieces of data making up 61 or 18). The correction factor $(GT)^2/N$ is the same (and always will be for this example).

$$SS \text{ geographic region} = (65^2 + 79^2)/14 - (144)^2/28 = 7.0$$

Next, we will determine the sum of squares for the political party × geographic region (actually stated as political party by geographic region) interaction. Here is the formula:

SS political party × geographic region = $(T^2_{Deast} + T^2_{Dwest} + T^2_{Reast} + T^2_{Rwest})/n - (GT)^2/N -$ SS political party – SS geographic region.

Let's break this apart to determine what is going on. The first part indicates the total (squared) for each cell and subtracting out the correction factor (one could call that the SS cells). Next, we subtract out the sum of squares of political party and geographic region (both are contained in the interaction). Why are we subtracting these sums of squares out?

When you examine the four cells, they really contain three things: the effect of political party, the effect of geographic region, and the interaction. Therefore, $(T^2_{Deast} + T^2_{Dwest} + T^2_{Reast} + T^2_{Rwest})/n$ is a combination of all three effects. Hence, if that's all there was to the sum of squares formula and we claimed statistical significance, then we wouldn't know if the effect was due to political party, geographic region, or the interaction. That's the reasoning behind why we subtract out the sums of squares political party and geographic region. By process of elimination, if we remove those effects, then the interaction is the remainder. Here are the numbers:

SS political party × geographic region
$(41^2 + 61^2 + 24^2 + 18^2)/7 - (144)^2/28 - 128.571 - 7.0 = 24.143$

The only issue here might be n. Once again, keep in mind that n is the number of pieces of data that make up 41 (Democrats, east), 61 (Democrats, west), etc. Hence, n is 7 in this case. If you took N (28) and divided by the number of groups (which is the multiplication of the levels of the independent variables – 2 × 2), then 28/4 = 7.

SS within = SS total – SS political party – SS geographic region –
SS political party × geographic region

If you remember from the one-way ANOVA, the SS within was calculated by taking the SS total – SS of the independent variable. The same logic holds true here. Take the SS total and subtract out all the other the other sum of squares to obtain the remainder (i.e., the within or error term). If you want to obtain the SS total and come back to the SS within, then that might be advantageous.

SS within = $237.429 - 128.571 - 7 - 24.143 = 77.714$

SS Total = $\Sigma\Sigma X^2/(GT)^2/N$

Similar to the one-way ANOVA, we begin by taking each piece of datum, squaring it, and then adding them. We do this for all 28 pieces of data. Hence, $8^2 + 5^2 + 4^2 + 7^2 + \cdots + 3^2 = 978$

SS total = $978 - (144)^2/28 = 237.429$

Now that we have our sums of squares, we now need to complete the summary table.

SUMMARY TABLE

The summary table for our 2×2 design is shown in Table 5.3:

Table 5.3 Summary Table for the two-way ANOVA

Source	Df	SS	MS	F	p
Political party	1	128.57	128.57	39.70	p < .01
Geographic region	1	7.00	7.00	2.16	p > .05
Political party × geographic region	1	24.143	24.143	7.45	p < .05
Within	24	77.714	3.238		
Total	27	237.429			

Let's take a look at each piece and refresh our memory a bit while learning a little something new. For political party, the degrees of freedom are $p - 1$ (political parties -1). There are two political parties $- 1$, so the df $= 1$. The mean square (MS) is obtained by dividing SS political party by df political party.

Likewise, for geographic region, the degrees of freedom are $r - 1$ (or geographic region -1). There are two geographic regions $- 1$, so the df $= 1$. Again the mean square (MS) is calculated by dividing SS geographic region by df geographic region.

The interaction, however, is a little different. The df here are multiplying $p - 1$ and $r - 1$ (or df political party × df geographic region). Hence, $1 \times 1 = 1$. The mean square is computed the same way (SS political party × geographic region/df political party × geographic region).

For the within, you obtained the SS within by taking SS total – SS political party – SS geographic region – SS political party × geographic region. Now apply the same concept for the df, that is df total – df political party– df geographic region – df political party × geographic region ($27 - 1 - 1 - 1 = 24$). It can also be obtained by taking $N - c$ (number of cells). The number of cells in a 2×2 design is 4 (multiply 2 by 2). Therefore, $28 - 4 = 24$. Moreover, you could also use $c(n - 1)$. The number of cells is 4 and n (the number of subjects in a group is 7). Therefore, $4 (7 - 1) = 24$. Incidentally, the df total is $N - 1$. If N is 28 (total number of pieces of data), then the df is 27.

CONCLUSIONS

Calculating the F ratios for each of the tested hypotheses is performed in the same way. Take the MS of the hypothesis and divide by the MS within. This is the same procedure we performed for the one-way ANOVA. For political party, take the MS political party (128.57) and divide by the MS within (3.238) and you'll get the F ratio, 39.7. Next, go to the F table on 1 and 24 degrees of freedom. The critical values are 4.26 (.05 level) and 7.82 (.01 level). Of course, our computed F ratio (39.7) is larger than both critical values; therefore, $p < .01$. If we examine the means, we find Democrats = 7.28, whereas Republicans = 3.00. The conclusion would be that Democrats had significantly higher satisfaction ratings than did Republicans.

For geographic region, the F ratio is not higher than either critical value; therefore, $p > .05$. The conclusion is that there is no statistically significant difference between the east (mean = 4.64) and west (mean = 5.64) geographic regions with regard to satisfaction ratings.

There is a statistically significant political party $\times$ geographic region interaction. This is the conclusion from the omnibus test. However, this conclusion is fairly ambiguous. Therefore, we must perform some additional testing in order to determine what is happening here. Whenever you have a statistically significant interaction, the first thing you should do is to plot it. After all, pictures are always worth 1000 words. There are three steps for plotting a two-way interaction.

1. Place the levels of an independent variable on the X-axis.

2. The values of the cell means are placed on the Y-axis.

3. Plot a single curve for each level of the other independent variable.

Figure 5.1 Plot of the political party $\times$ geographic region interaction

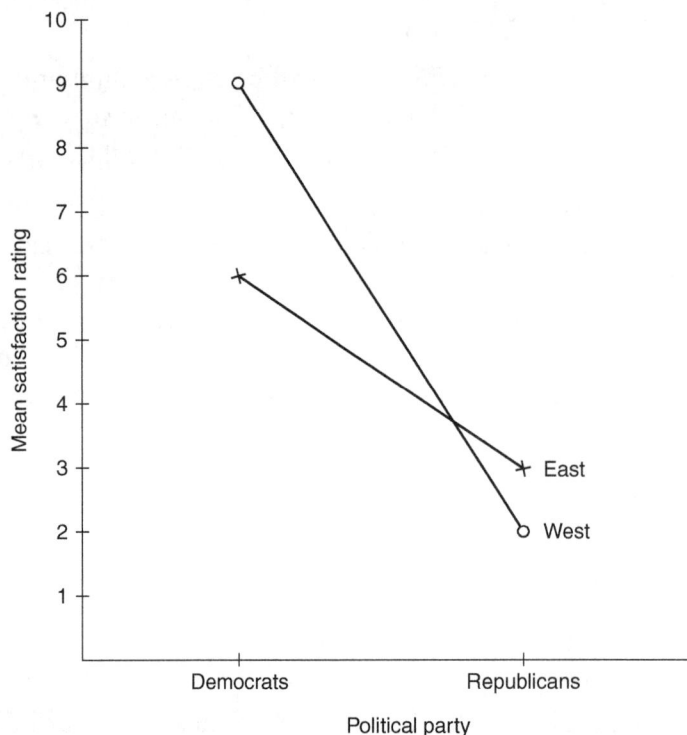

Figure 5.2 Plot of the political party × geographic region interaction

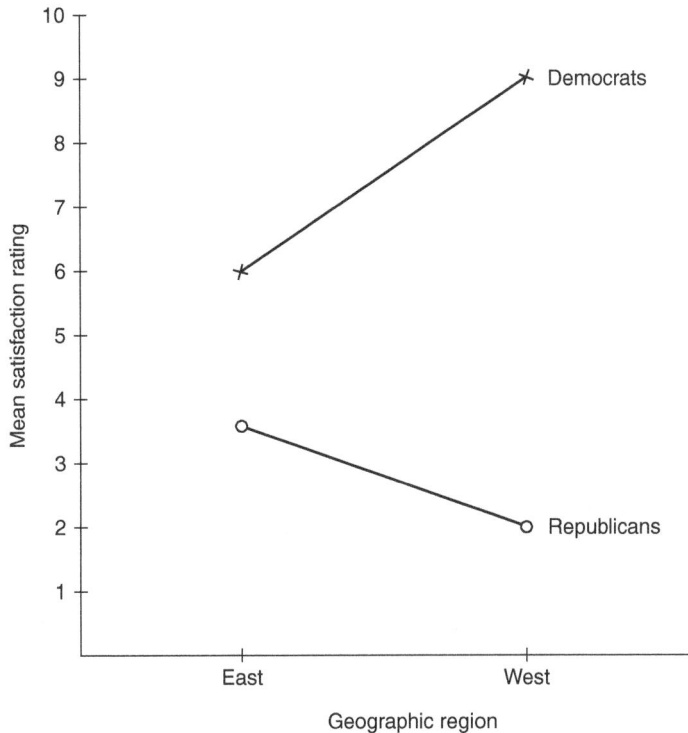

Let's briefly go through the steps. In Figure 5.1, the levels of political party are placed on the X-axis and we label the X-axis as political party. On the Y-axis, the label is mean satisfaction rating. The means are given in our two-way ANOVA example. Next, we plot the mean for Democrats East (5.857) and the mean for Democrats West (8.714) and connect the dots (we used x's for the East). Likewise, we plot the means for the Republicans East (3.428) and the Republicans West (2.571) and connect the dots (we used o's for the West). You could either make a legend signifying East and West or provide the label next to the line within the figure. This is usually the decision of the journals. In Figure 5.2, the same steps were followed.

Whenever you plot a 2-way ANOVA, it is always a good idea to put the independent variable with the most levels on the X-axis. Hence, you will be plotting fewer lines making the graph easier to interpret. In this case, it is a moot point, because both variables have two levels. Graphing the other way (more lines and fewer levels of the independent variable on the X-axis) is not technically incorrect, but it might be a bit more difficult to interpret. Let's demonstrate this with what I humorously call the "wood design", or 2 × 4. Suppose we have the following data as illustrated in Table 5.4 (the means for each cell are given):

Table 5.4 Hypothetical example of means for a 2 × 4 independent groups design

	B1	*B2*	*B3*	*B4*
A1	3	4	5	6
A2	6	5	4	3

Figure 5.3 A plot of the 2 × 4 interaction

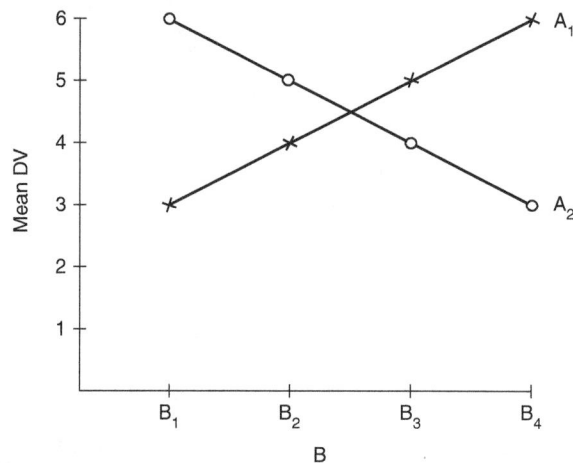

Figure 5.4 A plot of the 2 × 4 interaction

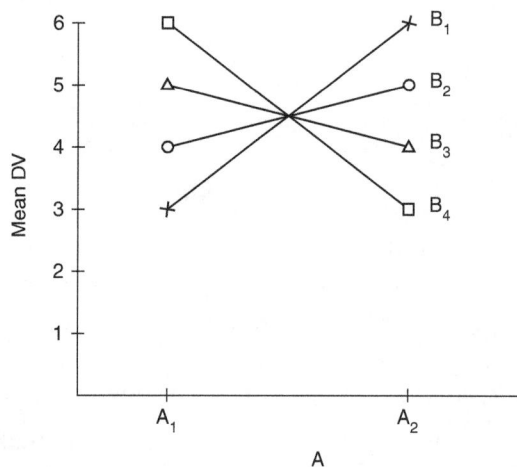

I hope you'll agree that Figure 5.3 is less cluttered than Figure 5.4, thereby making the former figure easier to interpret.

Moreover, you will notice that all four figures (Figures 5.1 – 5.4) have lines that are not parallel. That is the graphic way of determining an interaction. The lines do not have to cross (as illustrated in Figure 5.2), they just need not be parallel. These lines are called simple effects. A **simple effect** is the effect of an independent variable at a single level of the other independent variable. If we take you back to Figure 5.1, you will see the simple effect of political party at east and the simple effect of political party at west. Likewise, in Figure 5.2, there is the simple effect of geographic region at Democrats and the simple effect of geographic region at Republicans. Hence, in a 2 × 2 design, there are four simple effects. How many simple effects are there in the 2 × 4 design (without looking)? If you count the lines, then there are 6 simple effects. To determine the number of simple effects in a two-way interaction, count the lines or add the levels of each variable together. Hence, in a 2 × 4 design, it would be 2 + 4 = 6. In the 2 × 2 design, the number of simple effects is 4 (2 + 2). Thus, the number of simple effects in a two-way ANOVA is additive rather than multiplicative. Graphically, we can think of an interaction as a lack of parallelness of the simple effects.

Figure 5.5 Plot of the political party × geographic region interaction with the illustrated main effect of political party

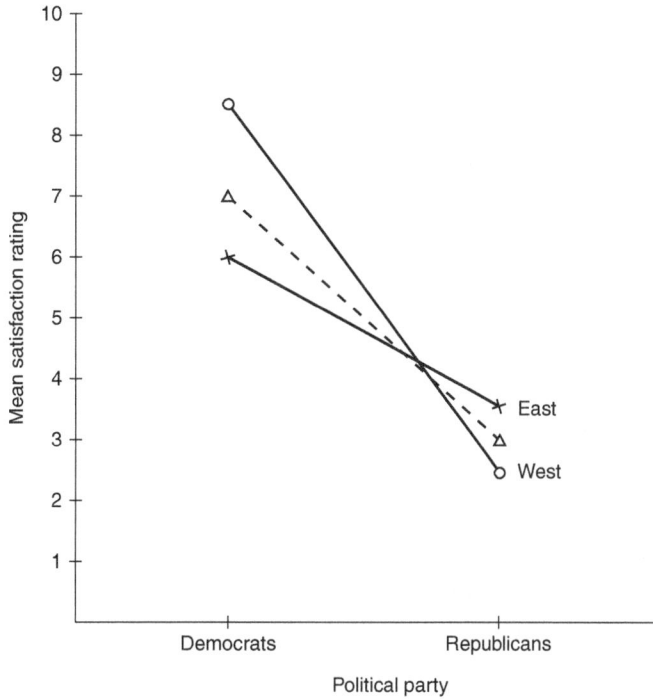

Figure 5.6 Plot of the political party × geographic region interaction with the illustrated main effect of geographic region

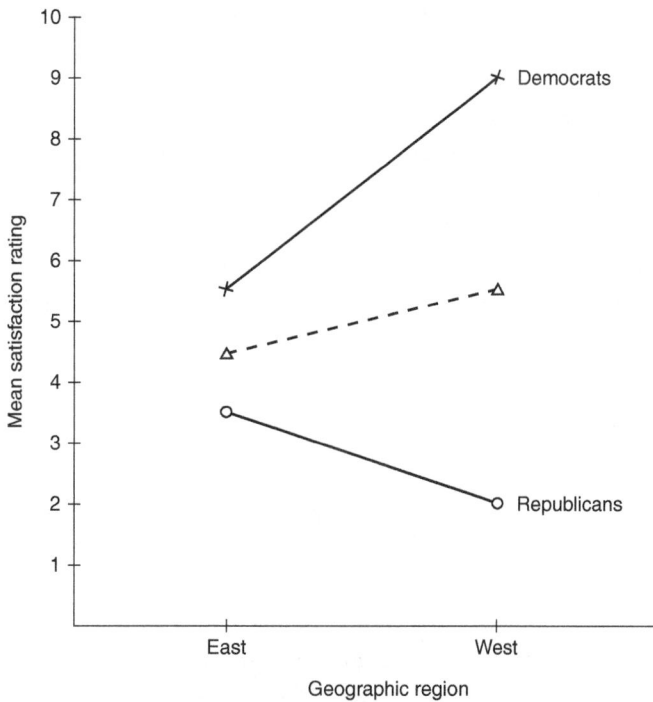

In Figure 5.5, suppose I take the average of Eastern Democrats and Western Democrats and plot a point. Likewise, if I do the same thing for Eastern Republicans and Western Republicans and plot a point and connect the line (as illustrated by the broken line), then that line would represent the main effect of political party. In Figure 5.6, suppose I take the average of Eastern Democrats and Eastern Republicans and plot a point. Likewise, if I do the same thing for Western Democrats and Western Republicans and plot a point and connect the line (as illustrated by the broken line), then that line would represent a main effect of geographic region. A **main effect**, therefore, is the average of the simple effects. However, a better definition is the overall difference among the levels of the independent variable tested. In this case, there are two main effects, one for political party and one for geographic region. The number of main effects is the number of independent variables that you have. If you examine the main effect in Figure 5.5, the slope of the line is fairly substantial. This indicates the statistically significant main effect of political party. However, in Figure 5.6, there is little to no slope of the line, thereby leading to the conclusion that there is no statistically significant main effect of geographic region. We could also say that we have a statistically significant main effect of political party but a nonsignificant main effect for geographic region (as evidenced by the p values in the summary table).

Although plotting the interaction is a nice start, we have provided neither any additional information nor conclusions to the study. In order to do that, we must conduct additional tests. Thus far, we addressed orthogonal comparisons, the Bonferroni test, and the Scheffe test. The next tests that we will discuss are a bit more versatile. These tests will work for both statistically significant main effects (more than two levels) and the interaction. So, let's say hello to range tests.

RANGE TESTS

As you will recall (I hope), the t-test examines the difference between group means. Of course, when you have multiple t-tests, the Type I error rate increases, thereby leading to false positives. **Range tests**, which examine all pairwise comparisons between means, are quite similar. However, they try to keep the Type I error rate down, while still having reasonable power. The original idea of range tests came from Gosset (who also developed the t-test). In fact, when he conceptualized the range distribution, he signed the paper "A Student." Henceforth, the distribution associated with the range test is now called the Studentized range distribution (in honor of Gosset's pseudonym). The **Studentized range distribution** is a distribution of a t-test run between the largest and smallest means from a set of k means. This can be shown empirically in the following way: Suppose I program the computer to give me three groups with five subjects per group. In this case, the mean of group 1 = 20 with a variance of 3, the mean of group 2 = 40 with a variance of 5; and the mean of group 3 = 50 with a variance of 6. Of course, one could compute the t-test between the largest and smallest means (groups 1 versus 3) via $(\overline{X}_1 - \overline{X}_2) / (\sqrt{(s_1^2 / n_1) + (s_2^2 / n_2)})$, which would be $(50 - 20) / (\sqrt{(3 / 5) + (6 / 5)})$.

The other formula for t is $(\overline{X}_1 - \overline{X}_2) / (\sqrt{(2 \times MS\ error) / n})$, which would be $(50 - 20) / (\sqrt{(2 \times 4.5) / 5})$. Suppose we obtain a different set of means for three groups with five subjects per group. The mean of group 1 = 40, variance = 5; the mean of group 2 = 60, with a variance of 8, and the mean of group 3 = 20 with a variance of 3. Again, a t-test between the largest and smallest means would be computed. If we do this simulation about 5000 times and plotted the value of the t-test on the x-axis with the frequency of occurrence on the y-axis, then we would have the Studentized range distribution.

Range tests should only be used after a statistically significant effect with more than two groups. They are more conservative and have less power than the omnibus F-test. Hence, if the F ratio is not statistically significant, then do not use range tests. We begin by examining some of the classic range tests. One of the positive points of many range tests is that the formulas are the same. However, the degrees of freedom change from test to test.

Tukey A

Named after its founder, John Tukey, this is also called the Tukey HSD (honestly significant difference) or the Tukey A (Tukey, 1953). Let's go through the steps that are fairly similar across range tests. First, we order the means from smallest to largest as shown in Table 5.5.

Table 5.5 Order of the means for the 2×2 interaction and the differences between them

	2.571 Republicans West (RW)	3.428 Republicans East (RE)	5.857 Democrats East (DE)	8.714 Democrats West (DW)
RW	—	.857	3.286	6.143
RE		—	2.429	5.286
DE			—	2.857
DW				—

You will notice that there are $p(p - 1)/2$ comparisons in which p represents the number of means. There are $4(4 - 1)/2$ or 6 possible pairwise hypotheses to be tested. The numbers contained in the table represent the differences between means. For example, for the comparison of Democrats West (DW) and Republicans West (RW), the difference between 8.714 and 2.571 is 6.143 (the value in the table). We did that for all possible comparisons.

Second, we need to compute a new statistic called q (the range test statistic). The formula is as follows:

$$q = (\overline{X}_i - \overline{X}_j) / (\sqrt{MS \text{ error} / n})$$

The table contains each numerator for each q value for testing the difference between means. For testing Democrats from the West against Republicans from the West, the $q = (6.143) / (\sqrt{3.238 / 7}) = 9.03$.

To determine whether this q value is statistically significant, use the total number of means (4) and the degrees of freedom error (24) to obtain the critical values in the q table. You will read this table similar to reading an F table. The top represents the numerator degrees of freedom (e.g., number of means), whereas the side represents the denominator degrees of freedom (e.g., from the df within in the two-way ANOVA). The critical values are 3.90 (.05) and 4.91 (.01) on 4 and 24 degrees of freedom. Once again, we ask the same question that we have been asking all along, "Is the q value that we compute higher than the critical values in the q table?" In this case, the answer is yes. 9.03 is larger than either 3.90 or 4.91 (p < .01). Hence, the conclusion is that Democrats who live in the West have significantly higher political party satisfaction scores than Republicans who live in the West.

If we are doing this by hand, we would move to the next largest difference (DW and RE). If this difference is not statistically significant, then none of the rest of the comparisons will be either via this method. For the comparison of DW and RE, $q = (5.286) / (\sqrt{3.238 / 7}) = 7.77$. This q value is higher than both critical values in the table, so it is statistically significant at p < .01. The conclusion is Democrats who live in the West have significantly higher satisfaction scores with their party than the Republicans who live in the East.

The third comparison examines DE with RW: $q = (3.286) / (\sqrt{3.238 / 7}) = 4.83$. This q value is higher than the first critical value in the table (3.90), so it is statistically significant at p < .05. The conclusion is Democrats from the East have significantly higher satisfaction scores with their party than the Republicans from the West.

The fourth comparison concerns DW and DE: $q = (2.857) / (\sqrt{3.238 / 7}) = 4.24$. This q value is higher than the first critical value in the table (3.90), so it is statistically significant at p < .05. The conclusion is that Democrats who live in the West have significantly higher satisfaction scores with their party than the Democrats who live in the East.

The fifth comparison is DE and RE: $q = (2.429) / (\sqrt{3.238 / 7}) = 3.57$. This q value is lower than either of the critical values in the table, so it is not statistically significant (p > .05). The conclusion is that there is no statistically significant difference in party satisfaction scores between Democrats who live in the East and Republicans who live in the East.

The final comparison contains RE and RW: $q = (.857) / (\sqrt{3.238 / 7}) = 1.26$. This q value is lower than either of the critical values in the table, once again, it is nonsignificant (p > .05). The conclusion is that there is no statistically significant difference in party satisfaction scores between Republicans who live in the East and Republicans who live in the West.

This can get quite laborious when there are many comparisons. Even with six comparisons, the conclusions can be a bit tedious. My suggestion is to combine the conclusions, so instead of having six separate sentences, they can be condensed to a more manageable number. Here is one possibility.

Democrats who live in the West have significantly higher party satisfaction scores than Republicans who live in the West, Republicans who live in the East, ps < .01, and Democrats who live in the East, p < .05. Democrats who live in the East have significantly higher party satisfaction scores than Republicans who live in the West, p < .05, but there was no statistically significant difference in party satisfaction scores from Republicans who live in the East, p > .05. Finally, there was no statistically significant difference between Republicans from the West and Republicans from the East in terms of their party satisfaction scores, p > .05.

So, let's review the steps for the Tukey A:

1. Order the means from smallest to largest

2. Compute q for each comparison

3. Compare each q with the critical values of q on the number of means and df error.

One of the problems with the Tukey A is that it is conservative. This means that you might not get as much power as you would like and perhaps deserve. Therefore, there is another test which will enhance your power and that is the Student Newman–Keuls (SNK).

Student Newman–Keuls

This range test developed by two different researchers in two separate papers (Newman, 1939; Keuls, 1952) purports to have greater power than the Tukey A. Because both researchers developed the test, that's the basis for Newman–Keuls (Newman probably received top billing as he published first). It is also called Student given that it is evaluated on the Studentized range distribution (or q tables). Let's examine this procedure using the same basic table (Table 5.5).

Table 5.5 Order of the means for the 2×2 interaction along with their differences

	2.571 Republicans West (RW)	3.428 Republicans East (RE)	5.857 Democrats East (DE)	8.714 Democrats West (DW)
RW	—	.857	3.286	6.143
RE		—	2.429	5.286
DE			—	2.857
DW				—

You'll notice that we ordered the means from smallest to largest and we will calculate the q value in the same manner as with the Tukey A. The only difference is in terms of the critical values. Here, the critical values of q are obtained by the number of ordered means (or steps) and the df error. The key component here is the number of ordered means. In the first comparison, Democrats West and Republicans West, how many means or steps are there from one group to another? We count each mean as a step. Democrats West is 1, Democrats East is 2, Republicans East is 3, and Republicans West is 4. Therefore, this comparison would be evaluated on 4 and 24 degrees of freedom. This is exactly the same as it would be for the Tukey A. This is always true for the top right comparison. Obviously, the conclusion will be the same for both range tests.

The difference between the Tukey A and SNK procedures is plainly seen in the remainder of the hypotheses. If we examine Democrats from the West against Republicans from the East, then in terms of the number of means from one to another in magnitude, this would be 3 (Democrats West, Democrats East, and Republicans East; that is, how many steps does it take to go from Democrats West to Republicans East?). Likewise, if we examine the comparison of Democrats East to Republicans West, this would also be 3 steps (Democrats East is 1, Republicans East is 2, and Republicans West is 3). Looking at the q table, the critical values would be 3.53 (.05) and 4.55 (.01). Remember, for the Tukey A, the critical values were 3.90 (.05) and 4.91 (.01). Given that the critical values for the SNK are lower than that of the Tukey A, they will be easier to beat, thereby giving us a bit more power for these comparisons.

Finally, if we examine the comparisons for Republicans West and Republicans East; Republicans East and Democrats East; and Democrats West and Democrats East, these are all on two steps (How many means are there from point A to point B? You must count each mean including the ones that are being compared). The df error would still be 24. From the q table, the critical values are 2.92 (.05) and 3.96 (.01). Once again, this has more power than 3.90 (.05) and 4.91 (.01) from the Tukey A. Here are the conclusions for the study using the SNK procedure.

Democrats who live in the West have significantly higher party satisfaction scores than Republicans who live in the West, Republicans who live in the East, and Democrats who live in the East, ps < .01. Democrats who live in the East have significantly higher party satisfaction scores than Republicans who live in the West, p < .01, and Republicans who live in the East, p < .05. Finally, there was no statistically significant difference between Republicans from the West and Republicans from the East in terms of their party satisfaction scores, p > .05.

If you compare the conclusions, you will notice a couple of differences. First, the Democrats who live in the West had significantly higher party satisfaction scores than Democrats who live in the East (p < .01 compared to p < .05 with Tukey A). The second difference is that Democrats who live in the East had significantly higher party satisfaction scores than Republicans who live in the East, p < .05 as opposed to being nonsignificant with Tukey A. Obviously, that last conclusion is a major turnaround.

You will also notice a pattern with the SNK as shown in Table 5.6. The top right comparison (in parenthesis; { }) is the number of steps (which is the same as the number of means, in this case only), the next diagonal (in curved brackets; [ ]) is the number of steps (number of means −1) and the final diagonal (in straight line brackets; | |) is the number of steps (number of means −2). You can never have less than 2 degrees of freedom for the numerator of the range test.

Table 5.6 Order of means and their differences for the 2 × 2 interaction along with the degrees of freedom

	2.571 Republicans West (RW)	3.428 Republicans East (RE)	5.857 Democrats East (DE)	8.714 Democrats West (DW)
RW	—	|.857| 2, 24	[3.286] 3, 24	{6.143} 4, 24
RE		—	|2.429| 2, 24	[5.286] 3, 24
DE			—	|2.857| 2, 24
DW				—

Recapping the steps for the SNK:

1. Order the means from smallest to largest

2. Compute q for each comparison

3. Compare each q with the critical values of q based on the number of ordered means (or number of steps) and df error.

The first two steps are the same for both Tukey A and SNK. The difference lies in the third step.

Although it is true that the SNK has more power and is still used today, Petrinovich and Hardyck (1967) found that this test has some liberal tendencies under various conditions. Of course, the Tukey A is a little conservative, so a compromise position is possible, known as the Tukey B.

Tukey B

This test (also called Tukey's Wholly Significant Difference Test; Tukey, 1953) tries to maintain Type I error rate a bit better than the SNK, yet it has more power than the Tukey A. In short, it takes the average of the critical values of the SNK and Tukey A in order to determine statistical significance.

In referring back to our example, the comparison containing Democrats from the West and Republicans from the West (the one with the largest difference) had critical values of 3.90 for Tukey A and for SNK at the .05 level and 4.91 for each at the .01 level. Hence, if we take the average of those critical values for Tukey A and SNK at the .05 level (3.90 + 3.90)/2 and for the .01 level (4.91 + 4.91)/2, obviously, the critical values for Tukey B will be the same. For the comparison with the largest difference between means (the top right), we can say that for Tukey A, SNK, and Tukey B, the qs, critical values, p values, and conclusions will be exactly the same.

If we examine Democrats from the West against Republicans from the East and the comparison of Democrats East with Republicans West, the critical values are 3.53 (.05) and 4.55 (.01) for the SNK, whereas they are 3.90 (.05) and 4.91 (.01) for the Tukey A. The Tukey B would have critical values of 3.715 (.05) and 4.73 (.01). These would be the average of the critical values (3.53 + 3.90)/2 and (4.55 + 4.91)/2. We would then use our q values to determine if they are higher than these new (average) critical values. Of course, if they are, then there is a statistically significant difference between those means.

Finally, if we examine the comparisons for Republicans West and Republicans East, Republicans East and Democrats East, and Democrats West and Democrats East, the critical values are 2.92 (.05) and 3.96 (.01) from the SNK. The Tukey A critical values are still 3.90 (.05) and 4.91 (.01). Hence, the critical values for the Tukey B would be 3.41 (.05) and 4.435 (.01), which are the averages. Here are the conclusions from the Tukey B.

Democrats who live in the West have significantly higher party satisfaction scores than Republicans who live in the West, Republicans who live in the East, and Democrats who live in the East, ps < .01. Democrats who live in the East have significantly higher party satisfaction scores than Republicans who live in the West and Republicans who live in the East, ps < .05. Finally, there was no statistically significant difference between Republicans from the West and Republicans from the East in terms of their party satisfaction scores, p > .05.

Although the conclusions were the same for the SNK and the Tukey B, you will notice that the significance level for the conclusion of Democrats who live in the East having significantly higher party satisfaction scores than Republicans who live in the West went from p < .01 with the SNK to p < .05 with the Tukey B. Here you will notice the slightly lower power.

Recapping the steps for the Tukey B:

1. Order the means from smallest to largest

2. Compute q for each comparison

3. Average the critical values of the Tukey A and SNK for each comparison.

Our next test focused truly on power. Does it control Type I error (keep it at the nominal level) and still have great power to boot? If so, then it is a marvelous test.

Duncan's Multiple Range Test

In Duncan's (1955) multiple range test, he focused on trying to obtain the greatest power possible. If you remember, one of the most powerful tests was orthogonal comparisons. In fact, he used the experimentwise error rate of orthogonal comparisons, namely, $1 - (1 - \alpha)^{g-1}$, which he called the protection level. Moreover, he also used the step-down procedure of the SNK which also allows for greater power, given that the critical values decrease with fewer steps. Finally, he developed his own critical values table in order to take these two major issues into account. Nevertheless, do you think that there is a problem with his test? If we examine all the comparisons from the range tests in terms of coefficients, then this would be illustrated in Table 5.7.

Table 5.7 All possible pairwise comparisons in terms of coefficients

RW	RE	DE	DW
+1	0	0	−1
+1	0	−1	0
+1	−1	0	0
0	+1	0	−1
0	+1	−1	0
0	0	+1	−1

These coefficients represent all six comparisons. For example, the first set (read across) would be comparing Republicans from the West with the Democrats from the West. To refresh your memory, each of these comparisons is valid (the sum of the coefficients equals 0); however, there is no independence. If you take the $\Sigma a_i b_i$ for the first two comparisons, then they would equal +1, not 0, thereby making the entire set of comparisons nonindependent. Moreover, setting the experimentwise error equal to orthogonal comparisons guarantees a higher Type I error rate. It was fine for orthogonal comparisons because they were independent, but not so for multiple t-tests or here. In fact, Petrinovich and Hardyck (1967) showed that this test was similar to performing multiple t-tests in terms of an increased Type I error rate and a lack of independence. Therefore, this test cannot be recommended for your use. As you see, there have been difficulties with each of the range tests discussed. Are there any tests that are at least "reasonable" in terms of Type I error and power?

Newer Range Tests

Given the problems associated with the SNK test as shown in Petrinovich and Hardyck's (1967) simulation, there have been a plethora of papers (primarily in the statistical literature) and procedures that have attempted to rectify the problems with the SNK. Such tests include the Ryan multiple comparison procedure, Welsch step-up, Welsch step-down, Shaffer-Ryan, and Peritz Q, just for starters. A number of these procedures (e.g., Peritz Q) are a bit difficult to compute by hand, so statisticians have created stand-alone programs (e.g., Martin & Toothaker, 1989) for performing these tests if they have not already been programmed into the standard statistical software packages. Much of the discussion concerning these tests is beyond our scope here. However, there is a fairly simple technique proposed by Hayter (1986) that purportedly has better Type I error properties than the Tukey or SNK tests and it also has reasonable power. This is called the **Fisher–Hayter** (FH) **test**. Here are the steps for the FH test.

1. Order the means from smallest to largest

2. Compute the q for each comparison

3. Use the number of means − 1 and the df error to obtain the critical value of q.

In our example, all comparisons would be evaluated on 3 and 24 degrees of freedom and the critical values would correspond to 3.53 (.05) and 4.55 (.01). Again, we ask the question if the q value that we calculate is larger than the critical value(s) in the q table.

Therefore, the conclusions for our study would be the following:

Democrats who live in the West have significantly higher party satisfaction scores than Republicans who live in the West, Republicans who live in the East, and Democrats who live in the East, ps < .01. Democrats who live in the East have significantly higher party satisfaction scores than Republicans who live in the West, and Republicans who live in the East, ps < .05. Finally, there was no statistically significant difference between Republicans from the West and Republicans from the East in terms of their party satisfaction scores, p > .05.

Because this test is evaluated on 3 and 24 degrees of freedom as compared to the 4 and 24 degrees of freedom for the Tukey A, SNK, and Tukey B, for the largest step comparison (DW versus RW), the critical value will be slightly less, thereby providing a bit more power. For the intermediate step comparisons, both SNK and FH evaluate each on 3 and 24 degrees of freedom, so there would be no difference in power here. For the smaller step comparisons, the SNK evaluates each on 2 and 24 degrees of freedom, whereas the FH continues to evaluate the comparisons on 3 and 24 degrees of freedom. Given that there are three comparisons here, the power would be slightly lower for each as compared to the SNK.

Final Thoughts on Range Tests

As range tests perform all pairwise comparisons between means, similar to multiple t-tests, the only difference is that the Type I error rates are more controlled for the range tests as compared to the t-test.

Hence, you would expect that there is a relationship between q and t. One formula for t is $(\overline{X}_1 - \overline{X}_2)/(\sqrt{(2 \times MS\ error)/n}$, whereas the formula for q is $(\overline{X}_1 - \overline{X}_2)/(\sqrt{MS\ error/n})$. The only difference between these two formulas is the $\sqrt{2}$. Let's put this to the test. If you examine the critical value of t on infinity degrees of freedom at the .05 level (1.96) and multiply by $\sqrt{2}$, then that will give you 2.77. Next, go to the q table on 2 and infinity degrees of freedom (it is 2 because we are dealing with two groups or two different means). That will also give you a critical value of 2.77. This relationship holds for any value of t. In a second example, take 10 degrees of freedom (critical value of t = 2.228 at the .05 level). The q value on 2 and 10 degrees of freedom at the .05 level is 3.15. For the .01 level, the critical value of t = 3.169, whereas the critical value of q is 4.48. So, the relationship is as follows: $q = t\sqrt{2}$ or $t = q/\sqrt{2}$.

When we are using range tests for an interaction, there are some comparisons that do not make a great deal of sense. In our example, comparing Democrats from the East to Republicans from the West, if there is a difference between these groups, then from where is this difference? Is it due to political party or geographic region? If you are not sure from where this difference comes, then this is confounding. Mathematically, the range test confounds both the main effects and interaction. However, it is used because of the ease of communication. The most mathematically ideal procedure that one can do is to orthogonally breakdown an interaction. Indeed, if I add up each orthogonal comparison (and I would have as many comparisons as there are df interaction), then they would equal the interaction sum of squares. Unfortunately, the communication leaves a bit to be desired (e.g., Is the change from A1B1 to A1B2 and A1B3 equal the change from A2B1 to A2B2 and A2B3?). This is pretty nebulous, so that is why it is not used. However, there is another way to examine cell differences in the interaction by confounding the interaction and only one main effect, yet still be able to communicate the results understandably. That would be the test of simple effects.

Test of Simple Effects

The test of simple effects combines some mathematical rigor with ease of interpretation. Like range tests, the test of simple effects would be used on an interaction after it is determined to be statistically significant. Unlike range tests, you cannot use the test of simple effects on anything but an interaction. Range tests are far more versatile, as they can be used on main effects and the interaction. Let's take a look at our example, perform the test of simple effects, and provide the conclusions.

If you remember, in a 2 × 2 design, there are four simple effects, so we will need to obtain the sum of squares and F values for each. The sums of squares are computed in the same way as the standard ANOVA. The first simple effect will be the effect of Democrats at geographic region.

SS Democrats @geographic region = $(T_{DE}^2 + T_{DW}^2)/n - (GT^2/N)$
    $(41^2 + 61^2)/7 - (102^2/14) = 28.57142$

SS Republicans @ geographic region = $(T_{RE}^2 + T_{RW}^2)/n - (GT^2/N)$
$(24^2 + 18^2)/7 - (42^2/14) = 2.57142$

SS East @ political party = $(T_{DE}^2 + T_{RE}^2)/n - (GT^2/N)$
$(41^2 + 24^2)/7 - (65^2/14) = 20.64285$

SS West @ political party = $(T_{DW}^2 + T_{RW}^2)/n - (GT^2/N)$
$(61^2 + 18^2)/7 - (79^2/14) = 132.07142$

The ANOVA table from our example with the tests of simple effects added after the interaction is shown in Table 5.8.

Table 5.8 Summary table for satisfaction ratings for the 2 × 2 ANOVA inclusive of the tests of simple effects

Source	Df	SS	MS	F	P
Political party	1	128.57	128.57	39.70	p < .01
Geographic region	1	7.00	7.00	2.16	p > .05
Political party × geographic region	1	24.143	24.143	7.45	p < .05
Democrats @ geographic region	1	28.57	28.57	8.82	p < .01
Republicans @ geographic region	1	2.57	2.57	<1	p > .05
East @ political party	1	20.64	20.64	6.37	p < 05
West @ political party	1	132.07	132.07	40.78	p < .01
Within	24	77.714	3.238		
Total	27	237.429			

Each simple effect is on 1 df because there were only two groups each (e.g., Democrats West versus Democrats East). If the simple effect contained Democrats from the East, West, North, and South, then that simple effect would have 3 df (or g − 1). The mean squares of each simple effect were divided by the mean square within in order to obtain F. Moreover, each simple effect was evaluated on 1 and 24 degrees of freedom in the F table to determine statistical significance. The conclusions are as follows:

Democrats who live in the West have significantly higher party satisfaction scores than Democrats who live in the East, p < .01. Moreover, Democrats who live in the West have significantly higher party satisfaction scores than Republicans who live in the West, p < .01. Democrats from the East have significantly higher party satisfaction than Republicans from the East, $p < .05$. Finally, there was no statistically significant difference between Republicans from the West and Republicans from the East in terms of their party satisfaction scores, $p > .05$.

Notice that there are fewer conclusions here than with the range tests as we have eliminated comparisons of groups with no levels of an independent variable in common (e.g., Democrats from the East versus Republicans from the West). Now, let's look at the confounding. If you add up the sums of squares for Democrats @ geographic region and Republicans @ geographic region, then you will obtain 31.14 (28.57 + 2.57). If you add up the sums of squares for the political party × geographic region interaction and the geographic region, you will also obtain 31.14 (24.14 + 7). Therefore, this would indicate the confounding of geographic region with the interaction. Likewise, if you add the sums of squares of East @political party with West @political party, then you will obtain 152.71 (20.64 + 132.07). If you add the sums of squares of the political party × geographic region interaction with political party (24.143 + 128.57), then you will also obtain 152.71. This demonstrates the confounding of political party with the interaction.

Although this was a 2×2 interaction, I earlier referred to the simple effect of Democrats from the East, West, North, and South. If that simple effect was statistically significant, then you would need to perform subsequent range tests (e.g., FH) on those four means. The statistically significant simple effect would indicate that there was a difference among the means, but we would not know where. This is similar to an omnibus F-test with more than two groups. In order to help you determine when to use subsequent tests and what to use, the following general rules of thumb will hopefully be helpful to you.

To Subsequent Test or Not to Subsequent Test

General Rules of Thumb

1. If you have only two levels of an independent variable, then do not subsequent test. If there is a statistically significant difference ($p < .05$ or $p < .01$), then check the means to find out which one is larger. Then you can state that group A is significantly higher than group B on the dependent variable. If there is no statistically significant difference ($p > .05$), then you can state that there is no statistically significant difference between groups A and B on the dependent variable.

2. If you have more than two levels of a statistically significant independent variable:

   a. If it's statistically significant, then do any of the subsequent tests we've covered (orthogonal comparisons, the Bonferroni test, the Scheffe test, or range tests). For the omnibus test, you would state that there is a statistically significant difference among the levels of the independent variable with regard to the dependent variable.

   b. If it's not significant, then you could still perform orthogonal comparisons or the Bonferroni tests (provided equal n's – it gets a little tricky with unequal n's). You can still do those tests because your hypotheses were developed a priori (before the data are collected). You will have problems obtaining power with the Bonferroni test if you have a bundle of comparisons. If the overall F isn't significant, then don't bother with the Scheffe or range tests.

3. If you have a statistically significant interaction, then you can

   a. Do tests of simple effects (and follow-up range tests if the simple effect is statistically significant and has more than two cells)

   b. Do range tests

4. If the interaction is not statistically significant, then no subsequent test is performed.

Two-way ANOVA (A × B)

Between or Independent Groups Design Class Example

An expert in basic learning was interested in examining the effects of teaching methods (In-Class versus Online) and study habit (massed versus distributed practice) on a statistics test. There were five subjects in each cell.

Here is the hypothetical data set.

Subject	In-Class		Online	
	Massed	Distributed	Massed	Distributed
1	43	58	68	61
2	36	51	77	49
3	47	46	72	55
4	39	49	79	52
5	45	57	76	51
Total	210	261	372	268
Mean	42.0	52.2	74.4	53.6

Grand Total = 210 + 261 + 372 + 268 = 1111

1. What are the null hypotheses?

2. What are the independent and dependent variables?

3. Know the summary table and be able to draw the interaction.

4. Know how to compute range tests. What are the differences among the types of range tests?

5. Draw appropriate conclusions.

Homework 4

Two-way ANOVA (A × B design)

A school psychologist was interested in examining if anxiety had an effect on academic performance that was measured by scores on a reading comprehension test. Moreover, she was concerned with sex differences and any interaction between these two variables.

Do the ANOVA and complete the summary table. If the independent variable with three levels is statistically significant, then perform a subsequent test.

Regardless of whether the interaction is statistically significant or not, plot the two-way interaction and perform the Tukey A and SNK range tests. Draw the appropriate conclusions.

Here is the hypothetical data set.

	Anxiety		
	Low	Moderate	High
Males	21	59	25
	25	56	17
	27	55	26
	28	54	19
Females	70	17	51
	76	20	52
	77	30	53
	75	35	54

CHAPTER 6

Three-Way ANOVA—Between or Independent Groups Design (A × B × C)

Suppose a nursing researcher is interested in determining differences in the body mass index (BMI) of individuals divided equally by sex, age group (old is over 65 and young is under 65), and whether they own their house or rent an apartment. If you are not familiar with BMI, then here is a brief synopsis: under 18.5 (underweight), 18.5–24.9 (normal), 25–29.9 (overweight), 30+ (obese). The results of each person's BMI are provided below in Table 6.1.

Table 6.1 Hypothetical data indicating each person's BMI as a function of sex, age group, and dwelling.

	Male Old Own	Male Old Rent	Male Young Own	Male Young Rent	Total
	28	17	26	30	
	20	18	22	31	
	24	21	28	34	
	22	20	29	33	
Total	94	76	105	128	403
Mean	23.5	19	26.25	32	

	Female Old Own	Female Old Rent	Female Young Own	Female Young Rent	Total
	23	27	36	20	
	19	19	32	19	
	34	25	25	18	
	32	26	27	28	
Total	108	97	120	85	410
Mean	27	24.25	30	21.25	

NULL HYPOTHESES

We begin by determining the seven null hypotheses tested.

1. $\mu_M = \mu_F$
   There is no statistically significant difference between the population means of the males and females on their BMIs.

2. $\mu_O = \mu_Y$
   There is no statistically significant difference between the population means of the older and younger participants with regard to their BMIs.

3. $\mu_O = \mu_R$
   There is no statistically significant difference between the population means of owners and renters with regard to their BMIs.

4. There is no sex × age interaction in the population.

5. There is no sex × dwelling interaction in the population.

6. There is no age × dwelling interaction in the population.

7. There is no sex × age × dwelling interaction in the population.

SUMS OF SQUARES

Let us calculate the sums of squares for prospective inclusion into the summary table.

$$SS\ sex = (T^2_M + T^2_F)/n - (GT)^2/N$$

In order to obtain the totals, you must add over the age and dwelling. For males, the totals would be 94 (from the old who own) + 76 (from the old who rent) + 105 (from the young who own) + 128 (from the young who rent) for a total of 403. Likewise, for females, the totals would be 108 (from the old who own) + 97 (from the old who rent) + 120 (from the young who own) + 85 (from the young who rent) for a total of 410. The n is the number of pieces of data needed to obtain the 403 or 410. In this case, n = 16. There were 16 individual pieces of data making up 403 and 16 individual pieces of data making up 410. The grand total (GT) is 403 + 410 or 813. N is the total number of pieces of data, which are 32. Again, it could also be obtained by taking the number of the levels of the IV (2) times the number of pieces of data making up the 403 or 410, namely, 16. Thus, 2 × 16 = 32. Once again, you could also take N/the number of levels of the IV in order to obtain n (32/2 = 16).

$$SS\ sex = (403^2 + 410^2)/16 - (813)^2/32 = 1.531$$

We perform the same task for obtaining the SS age.

$$SS\ age = (T^2_{old} + T^2_{young})/n - (GT)^2/N.$$

We obtain the totals here by adding over sex and dwelling. For the old folks, we take the totals from males who own (94), males who rent (76), females who own (108), and females who rent (97) and add them together for a total of 375. Likewise, we perform the same task for the young folks: males who own (105), males who rent (128), females who own (120), and females who rent (85), for a total of 438. The n is still 16 (as there are 16 pieces of data making up 375 or 438).

$$SS\ age = (375^2 + 438^2)/16 - (813)^2/32 = 124.031$$

Likewise, we perform the same procedure for dwelling by adding over sex and age.

$$SS\ dwelling = (T^2_{own} + T^2_{rent})/n - (GT)^2/N.$$

Once again, we will go through the same procedure for obtaining the totals. For owning, add the totals of males who are old (94), males who are young (105), females who are old (108), and females who are young (120) for a total of 427. For renting, add the totals for males who are old (76), males who are young (128), females who are old (97) and females who are young (85) for a total of 386. There are 16 pieces of data that made up the totals of owning and renting.

$$SS\ dwelling = (427^2 + 386^2)/16 - (813)^2/32 = 52.531$$

Next, we will determine the sum of squares for the sex × age interaction. When you have a smaller interaction (two-way) contained within a larger design (three-way), my suggestion is to make a table showing each level of sex and age and indicating their totals (i.e., adding over dwelling).

	Males	*Females*
Old	170	205
Young	233	205

In order to obtain the total for old males, add the total for males who are old and own (94) with the total for males who are old and rent (76) and that equals 170. Next, for obtaining the total for old females, add the total for females who are old and own (108) with the total for females who are old and rent (97) and that equals 205. Next, for obtaining the total for young males, add the total for males who are young and own (105) with the total for males who are young and rent (128) for a total of 233. Finally, for obtaining the total for young females, add the total for females who are young and own (120) with the total for females who are young and rent (85) for a total of 205.

Here is the formula:

$$SS\ sex \times age = (T^2_{Mold} + T^2_{Fold} + T^2_{Myoung} + T^2_{Fyoung})/n - (GT)^2/N - SS\ sex - SS\ age.$$

This is analogous as to how we calculated the two-way interaction from our last design.

Here are the numbers:

$$SS\ sex \times age\ (170^2 + 205^2 + 233^2 + 205^2)/8 - (813)^2/32 - 1.531 - 124.031 = 124.031$$

The only issue here might be n. Once again, keep in mind that n is the number of pieces of data that make up 170 (males, old), 205 (females, old), etc. Hence, n is 8. If you took N (32) and divided by the number of groups or cells (which is the multiplication of the levels of the independent variables – 2 × 2), then 32/4 = 8. Incidentally, it is a coincidence that the SS sex × age equaled the SS age. More often than not, this phenomenon will not occur.

Next, we will determine the sum of squares for the sex × dwelling interaction. Once again, because this is a smaller interaction contained within a larger design, it is suggested to make a table showing each level of sex and dwelling and indicating their totals (i.e., adding over age).

	Males	*Females*
Own	199	228
Rent	204	182

In order to obtain the total for males who own, add the total for males who are old and own (94) with the total for males who are young and own (105) and that equals 199. Next, for obtaining the total for females who own, add the total for females who are old and own (108) with the total for females who are young and own (120) and that equals 228. Next, for obtaining the total for males who rent, add the total for males who are old and rent (76) with the total for males who are young and rent (128) for a total of 204. Finally, for obtaining the total for females who rent, add the total for females who are old and rent (97) with the total for females who are young and rent (85) for a total of 182.

Here is the formula:

$$SS\ sex \times dwelling = (T^2_{Mown} + T^2_{Fown} + T^2_{Mrent} + T^2_{Frent})/n - (GT)^2/N - SS\ sex - SS\ dwelling.$$

Here are the numbers:

$$SS\ sex \times dwelling\ (199^2 + 228^2 + 204^2 + 182^2)/8 - (813)^2/32 - 1.531 - 52.531 = 81.281$$

Let's have another look at n. Keep in mind that n is the number of pieces of data that make up 199 (males, own), 228 (females, own), etc. Once again, n is 8. If you took N (32) and divided by the number of groups or cells, then 32/4 = 8.

Our final two-way interaction is age × dwelling. Make a table showing each level of age and dwelling and indicating their totals (i.e., adding over sex).

	Old	*Young*
Own	202	225
Rent	173	213

For the sake of completeness, let's go over the totals. In order to obtain the total for old people who own, add the total for males who are old and own (94) with the total for females who are old and own (108) and that equals 202. Next, for obtaining the total for young people who own, add the total for males who are young and own (105) with the total for females who are young and own (120) and that equals 225. Next, for obtaining the total for old people who rent, add the total for males who are old and rent (76) with the total for females who are old and rent (97) for a total of 173. Finally, for obtaining the total for young people who rent, add the total for males who are young and rent (128) with the total for females who are young and rent (85) for a total of 213.

Here is the formula:

$$\text{SS age} \times \text{dwelling} = (T^2_{\text{Oown}} + T^2_{\text{Yown}} + T^2_{\text{Orent}} + T^2_{\text{Yrent}})/n - (GT)^2/N - \text{SS age} - \text{SS dwelling}.$$

Here are the numbers:

$$\text{SS age} \times \text{dwelling} (202^2 + 225^2 + 173^2 + 213^2)/8 - (813)^2/32 - 124.031 - 52.531 = 9.031$$

Keep in mind that n is the number of pieces of data that make up 202 (old, own), 225 (young, own), etc. Once again, n is 8. If you took N (32) and divided by the number of groups or cells, then $32/4 = 8$.

Here is the formula for the three-way interaction:

$$\text{SS sex} \times \text{age} \times \text{dwelling} = (T^2_{\text{Moldown}} + T^2_{\text{Myoungown}} + T^2_{\text{Moldrent}} + T^2_{\text{Myoungrent}} + T^2_{\text{Foldown}} + T^2_{\text{Fyoungown}} + T^2_{\text{Foldrent}} + T^2_{\text{Fyoungrent}})/n - (GT)^2/N - \text{SS sex} - \text{SS age} - \text{SS dwelling} - \text{SS sex} \times \text{age} - \text{SS sex} \times \text{dwelling} - \text{SS age} \times \text{dwelling}$$

Now, for the numbers,

$$\text{SS sex} \times \text{age} \times \text{dwelling} = (94^2 + 76^2 + 105^2 + 128^2 + 108^2 + 97^2 + 120^2 + 85^2)/4 - (813^2/32) - 1.531 - 124.031 - 52.531 - 124.031 - 81.281 - 9.031 = 132.031.$$

The n is obtained by taking the total N (32) and divided by the number of cells (8). There are 8 cells in a 2 × 2 × 2 design. Hence, $32/8 = 4$. Of course, the other way to approach it is how many pieces of data made up the totals within each cell, 94, 76, etc. That would also be 4. In this case, you need to subtract out all sums of squares that are contained in the three-way interaction. This includes each main effect and all possible two-way interactions.

For the sum of squares within,

SS within = SS total − SS sex − SS age − SS dwelling − SS sex × age − SS sex × dwelling − SS age × dwelling − SS sex × age × dwelling

If you remember from the two-way ANOVA, the SS within was obtained by taking the SS total − SS of the independent variables − the sum of squares of the interaction. The same logic holds true here. Take the SS total and subtract out all the other sum of squares to obtain the remainder (i.e., the within or error term). Therefore, if you want to obtain the SS total and then come back to the SS within, then that might be advantageous.

$$SS\ within = 937.719 - 1.531 - 124.031 - 52.531 - 124.031 - 81.281 - 9.031 - 132.031 = 413.25$$

$$SS\ total = \Sigma\Sigma\Sigma X^2 - (GT)^2/N$$

Similar to the one-way or two-way ANOVA, we begin by taking each piece of datum, squaring it, and then adding them. We do this for all 32 pieces of data. Hence, $28^2 + 20^2 + 24^2 + 22^2 + \cdots + 28^2 = 21,593$

$$SS\ total = 21,593 - (813)^2/32 = 937.719$$

Now that we have our sums of squares, we now need to complete the summary table.

SUMMARY TABLE

The summary table for our $2 \times 2 \times 2$ design is contained in Table 6.2:

Table 6.2 Summary table for the three-way ANOVA

Source	df	SS	MS	F	p
Sex	1	1.531	1.531	<1	p > .05
Age	1	124.031	124.031	7.20	p < .05
Dwelling	1	52.531	52.531	3.05	p > .05
Sex × age	1	124.031	124.031	7.20	p < .05
Sex × dwelling	1	81.281	81.281	4.72	p > .05
Age × dwelling	1	9.031	9.031	.525	p > .05
Sex × age × dwelling	1	132.031	132.031	7.668	p < .05
Within	24	413.25	17.219		
Total	31	937.719			

Just to rehash, to obtain the df for sex, age, and dwelling, take the number of levels of each independent variable and subtract 1. To obtain the degrees of freedom for each two-way interaction, multiply the degrees of freedom for each effect. Once again, for sex × age, the degrees of freedom are obtained by taking the df sex × df age ($1 \times 1 = 1$). The degrees of freedom for the three-way interaction are obtained in the same way (df sex × df age × df dwelling; $1 \times 1 \times 1 = 1$).

For the within, as you obtained the SS within by taking SS total − SS all above (SS sex − SS age − SS dwelling − SS sex × age − SS sex × dwelling − SS age × dwelling − SS sex × age × dwelling), again apply the same concept for the df, that is, df total − df all above (df sex − df age − df dwelling − df sex × age − df sex × dwelling − df age × dwelling − df sex × age × dwelling, or $31 - 1 - 1 - 1 - 1 - 1 - 1 - 1 = 24$). It can also be obtained by taking N − c (# of cells). Therefore, $32 - 8 = 24$. Moreover, you could also use $c(n - 1)$. The number of cells is 8 and n (the number of subjects in a group is 4). Therefore, $8 (4 - 1) = 24$. The df total for is 31 (total number of pieces of data − 1). You could also add df sex through df within (all eight df terms) and you would obtain 31.

CONCLUSIONS

The conclusions for the design are the following: Young folks have significantly higher BMIs than older folks (look at the means: young = 438/16 = 27.37; old = 375/16 = 23.43). There were no statistically significant differences in BMIs with regards to sexes or dwellings. There was a statistically significant sex × age interaction. You could plot the two-way interaction and perform either tests of simple effects or range tests. Although there were no statistically significant sex × dwelling or age × dwelling interactions, there was a statistically significant sex × age × dwelling interaction. Once again, the first task is to plot it. In this case, to plot a three-way interaction, plot a two-way interaction for each level of the third variable. As shown in Figure 6.1, we plotted the sex × age interaction for each level of dwelling (own or rent). You could also plot the sex × dwelling interaction for each level of age or plot the age × dwelling interaction for each level of sex. If you had a $2 \times 2 \times 4$ interaction, again you might want to plot the 2×2 interaction for each level of the third independent variable, or you could plot the 2×4 (with the four levels on the x-axis) for each level of the independent variable with two levels. In other words, the fewer the lines (or simple effects) shown, the easier the interpretation usually is.

In fact, if you wanted to plot a four-way interaction, you would plot the two-way interaction for each level of the third and fourth variable. For example, you could plot the sex × age interaction for each level of dwelling and city (e.g., Las Vegas and Reno). Therefore, one graph would have the sex × age interaction for Las Vegas and own, one graph would be the sex × age interaction for Las Vegas rent, one graph would be the sex × age interaction for Reno own, and one graph would be the sex × age interaction for Reno rent.

One could also perform range tests across all eight means (28 significance tests looking at all pairwise comparisons between means: for example, older males who own versus older males who rent; older males who own versus younger males who own) or tests of simple effects are also an option. If you examine Figure 6.1, you will notice that for those who own, the slopes of the males' and females' BMIs across ages are fairly similar. The statistical significance of the interaction term comes from the renters. Notice that for males, the BMIs increase from old to young, whereas for females, the BMIs decrease slightly from old to young.

Figure 6.1 Plot of the sex × age × dwelling interaction

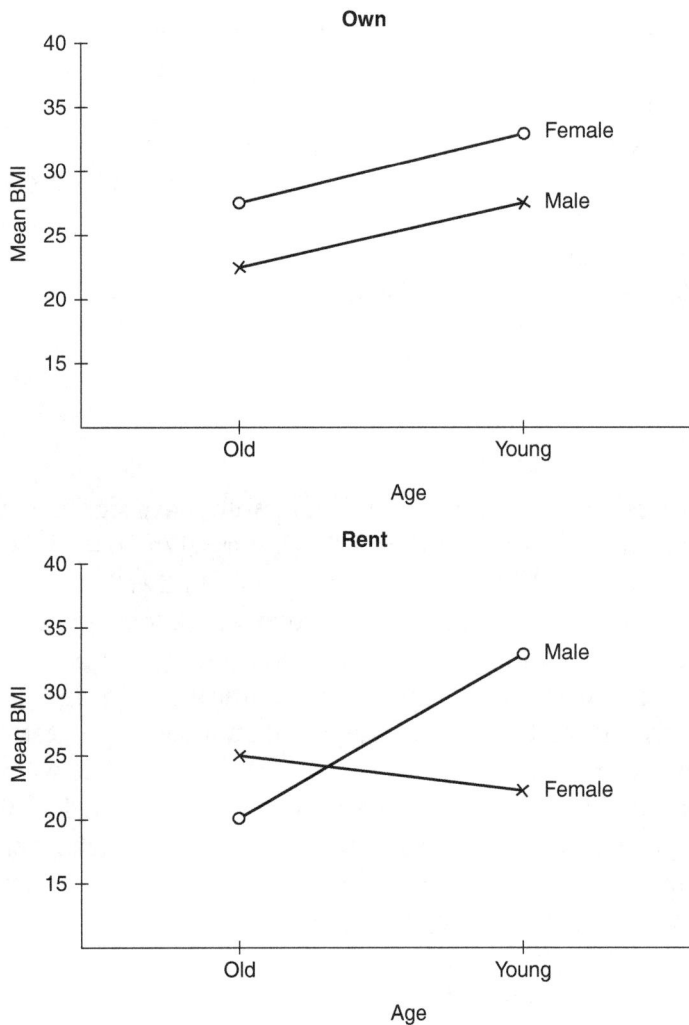

Three-way ANOVA

2 × 2 × 2 (Independent Groups Design) (A × B × C) Class Example

A Machiavellian Test (Christie & Geis, 1970) was administered to students at UNLV majoring in Psychology or Political Science. Five male undergraduates, five female undergraduates, five male graduate students, and five female graduate students participated. A Machiavellian personality refers to being conniving, deceptive, and amoral (e.g., using others for their own gain) on a regular basis without feeling guilt. Here is the hypothetical data set:

	Male *PSY Under*	*Male* *PSY Grad*	*Male* *PoliSci Under*	*Male* *PoliSci Grad*
	34	23	19	80
	23	38	80	93
	56	58	45	78
	61	70	34	69
	27	61	67	80
Total	201	250	245	400
Mean	40.2	50.0	49.0	80.0

	Female *PSY Under*	*Female* *PSY Grad*	*Female* *PoliSci Under*	*Female* *PoliSci Grad*
	39	50	40	80
	30	61	58	90
	20	48	40	87
	27	70	60	75
	35	75	68	92
Total	151	304	266	424
Mean	30.2	60.8	53.2	84.8

Grand Total = 201 + 250 + 245 + 400 + 151 + 304 + 266 + 424 = 2241

1. What are the null hypotheses?

2. What are the independent variables? What is the dependent variable?

3. Know how to obtain the summary table and the F ratios.

4. Know how to draw a three-way interaction.

5. Draw appropriate conclusions.

Homework 5

Three-way ANOVA (A × B × C)

Rob Manfred was interested in knowing if there were any significant differences in players' salaries as a function of position (infield, outfield, pitcher), division (east, central, west), and league (NL and AL—National League and American League). Show the summary table and if there are any statistically significant interactions, then plot them. Plot the three-way interaction even if it is not significant. Do subsequent tests where applicable. Show all work! Salary is provided in millions (e.g., 2.5 is 2.5 million). The hypothetical data is presented below.

NL East Infield	NL East Outfield	NL East Pitcher	NL Central Infield	NL Central Outfield	NL Central Pitcher	NL West Infield	NL West Outfield	NL West Pitcher
2.5	1.8	3.5	3.3	2.2	4.4	2.8	3.9	0.7
0.8	1.9	2.8	2.9	5.0	4.3	3.3	3.3	1.9
2.2	0.9	1.7	2.1	4.8	0.6	2.7	4.8	1.2
3.4	0.7	1.9	0.5	3.2	1.7	4.0	3.3	2.2
2.9	0.5	0.5	0.8	1.8	1.1	1.4	4.7	3.8

AL East Infield	AL East Outfield	AL East Pitcher	AL Central Infield	AL Central Outfield	AL Central Pitcher	AL West Infield	AL West Outfield	AL West Pitcher
4.5	4.1	0.5	2.8	5.7	2.4	3.8	0.9	0.7
2.3	0.8	5.8	2.4	1.0	1.5	0.1	0.4	4.9
0.2	2.1	2.7	2.8	1.4	3.6	0.7	1.2	0.2
1.9	3.6	4.7	0.3	3.3	4.7	0.2	0.9	2.6
0.5	0.5	3.1	0.2	1.3	2.1	0.7	0.2	0.8

CHAPTER 7

Unequal Sample Sizes
(unequal n)

Thus far, we have dealt with the sample sizes being equal in all conditions. Of course, if you sample from an Introduction to Psychology course, it is common to have twice as many females as males. Hence, if I were examining sex differences, then maybe I would have 400 females and 200 males. In fact, unequal sample sizes are more common than equal sample sizes if you are surveying a large class. However, if you are performing an experiment in neuroscience, which has more controls and lesser sample sizes, then it is more plausible that equal sample sizes would occur (barring attrition due to death). But, if you are conducting questionnaire or survey research, for example, then how does one deal with unequal n?

ONE-WAY ANOVA

In a one-way ANOVA, let's take the following example: Suppose that we have four groups of children from Elementary Schools in Henderson with the following disorders: ADHD, Autism, CAPD, and Williams Syndrome. These children would be tested on the number of blocks correctly placed on the Seguin Form Board Test. The results are shown in Table 7.1:

Table 7.1 Hypothetical results of the number of blocks correctly placed as a function of disorder for each of 17 children

ADHD	Autism	CAPD	Williams	
7	5	8	6	
5	2	6	4	
8	5	6		
6	4			
9	2			
3				
4				
42	18	20	10	Grand Total = 90

SUMS OF SQUARES

The sum of squares is performed in an analogous manner to the one-way ANOVA that we computed earlier. However, the only minor difference is that the sample size for each group is the denominator and it will change due to unequal n. When we had equal n, the sample size for each group was the denominator, but it was always the same for all groups.

SS disorder $= (T_{ADHD}^2/n_{ADHD} + T_{AUT}^2/n_{AUT} + T_{CAPD}^2/n_{CAPD} + T_{WILL}^2/n_{WILL}) - (GT)^2/N$

$\quad (42^2/7 + 18^2/5 + 20^2/3 + 10^2/2) - (90)^2/17 = 23.663$

SS within = SS total − SS disorder

$\quad 65.53 - 23.663 = 41.867$

SS total $= \Sigma X^2 - (GT)^2/N$

$\quad 542 - (90)^2/17 = 65.53$

Let's take a look at the summary table for unequal n as shown in Table 7.2.

Table 7.2 Summary table for the one-way ANOVA with unequal n

Source	df	SS	MS	F	p
Disorder	3	23.663	7.888	2.449	p > .05
Within	13	41.867	3.221		
Total	16	65.53			

The conclusion is that there is no statistically significant difference among the psychological disorders with regard to the number of blocks correctly placed on the Seguin Form Board. Mathematically, unequal n for a one-way ANOVA poses no real problem, albeit there is a slight loss of power.

ORTHOGONAL COMPARISONS

Suppose you wanted to perform orthogonal comparisons, then what might you do? Remember that the characteristics were $\Sigma a_i = 0$ (valid comparison) and $\Sigma a_i b_i = 0$ (independence) for equal sample sizes. However, the characteristics shift a bit for unequal n, namely, $\Sigma n_i a_i = 0$ (valid comparison) and $\Sigma n_i a_i b_i = 0$ (independence). Now, each group's sample size needs to be taken into account. Let's demonstrate this procedure with three groups as shown in Table 7.3.

Here we have three groups (A, B, and C) with different sample sizes, which are shown at the top.

Table 7.3 Hypothetical orthogonal coefficients for a one-way ANOVA with unequal n

n =	5	10	15
	A	B	C
	−1	−1	+1
	−2	+1	0

For the first comparison (groups A and B versus C), here is how you would determine a valid comparison:

$5(-1) + 10(-1) + 15(1) = (-15 + 15) = 0$. We multiplied n by the coefficient for the particular group and then summed them. If I draw a vertical line between the positive and negative coefficients (+1 and −1 – between groups B and C), and if I cannot cross the line, then C can no longer be compared. Therefore, the only comparison left is group A versus group B.

For the second comparison (group A versus group B),

$5(-2) + 10(1) + 15(0) = (-10 + 10 + 0) = 0$.

Both of these comparisons are valid. But, are they independent?

$5(-1)(-2) + 10(-1)(+1) + 15(+1)(0) = (10 + -10 + 0) = 0$

Yes, they certainly are. What we did here is to take the sample size for each group and multiply n by their coefficients for hypothesis 1 and hypothesis 2. Of course, I fixed the numbers a bit. If you had sample sizes that were 977, 40, and 152, then the coefficients would be more difficult to obtain. That is why it is rare (if at all), you will see orthogonal comparisons with unequal sample sizes. There is a way to perform subsequent tests with unequal n, but we will briefly address that a bit later.

TWO-WAY ANOVA

Suppose you have a two-way ANOVA, what happens when you have unequal n? Optimally, you would want a factorial design. A **factorial design** is one in which each level of each variable occurs equally often with each level of every other variable. You must have equal sample sizes. Thus far, our designs (in this book, in class, and for homework) have been factorial. To further elucidate, an example of a factorial design is shown in Table 7.4.

Table 7.4 A factorial design for a 2-way ANOVA as a function of sex and university

	UNLV	*UNR*
Male	n = 50	n = 50
Female	n = 50	n = 50

Here we would have 100 males and 100 females equally divided by university. Unfortunately, obtaining equal sample sizes per group is easier said than done. Suppose you had a case as shown in Table 7.5.

Table 7.5 A completely confounded design

	Saturday	*Sunday*
Jewish	n = 0	n = 40
Catholic	n = 40	n = 0

In this case, we could not sample Jewish folks on Saturday (at synagogue) and we could not sample Catholics on Sunday (at church). Suppose we found a difference between the Jewish on Sunday and Catholic on Saturday groups on a particular dependent variable? From where is the difference? Is the difference due to religion or is it due to day? We do not know. This is a case of complete confounding. **Complete confounding** is a design in which you do not know where the difference lies. Of course, no legitimate researcher strives for a completely confounded design, but it can happen by serendipity. For instance, in the neuroscience area, rats may die if too high a dosage of a drug is given (even though it may have been contracted in the grant).

Nevertheless, let's take an example and perform the standard two-way ANOVA in which we examine the number of hours texting on Saturday and Sunday for old (over 55) and young (under 55) people. The data are shown in Table 7.6.

Table 7.6 Hypothetical results for the number of hours texting as a function of age and day

	Saturday	*Sunday*	
Old	n = 5	n = 2	
	2,2,2,2,2	2,2	14
Young	n = 2	n = 5	
	5,5	5,5,5,5,5	35
Totals	20	29	49

For the old folks on Saturday cell, there were five individuals and each one texted for two hours each. For the old folks on Sunday cell, two individuals texted for two hours each. For the young folks on Saturday cell, two individuals texted for five hours each, whereas for the young folks on Sunday cell, five individuals texted for five hours each.

Let's compute the two-way ANOVA as we have done before.

SS age = $(14^2 + 35^2)/7 - (49^2/14) = 31.5$

SS day = $(20^2 + 29^2)/7 - (49^2/14) = 6.78571$

SS age × day = $(10^2/5 + 4^2/2 + 10^2/2 + 25^2/5) - (49^2/14) - 31.5 - 6.78571 = -6.78571$

Unequal n results in destroying the factorial nature of the design, and there is also a lack of robustness. As you can see, using the standard two-way ANOVA formula will not work, given the negative sum of squares for the age x day interaction. This does not mean that you always obtain a negative sum of squares for the interaction, but I "fudged" the numbers to prove a point. If we cannot use the standard ANOVA formulas, then what can we do to remedy the unequal n problem?

HOW TO GET RID OF UNEQUAL n

1. Randomly discard data

In the "olden" days, when computers were neither as sophisticated nor as prevalent, randomly discarding data to obtain equal n was appropriate. Then, the researcher used the standard ANOVA formulas for testing the hypotheses. Table 7.7 provides an example that would make sense for randomly discarding data.

Table 7.7 Example of unequal n in a 2 x 2 design with small discrepancy

	A1	*A2*
B1	n = 10	n = 9
B2	n = 9	n=9

In this case, one possibility would be to take a hat with the names or code numbers of the participants in the A1 B1 condition and randomly draw one. Another possibility would be to use a random number table to eliminate one of the participants (e.g., the one with the highest random number would be eliminated).

However, let's take a different example with large discrepancies as shown in Table 7.8.

Table 7.8 Example of unequal n in a 2 x 2 design with large discrepancies

	A1	*A2*
B1	n = 100	n = 900
B2	n = 90	n = 9

In this example, you would need to eliminate 91 cases from A1 B1, 891 cases from A2 B1, and 81 cases from A1 B2 in order to obtain equal n. That is 1063 out of 1099 cases (over 96%) that you would throw out. Obviously, this would not be a logical solution. Therefore, randomly discarding data is not feasible when the sample sizes are highly discrepant.

2. Yates substitution formula

Here is another "older" solution that was viable, provided the sample sizes were not too discrepant. Suppose we have the following as shown in Table 7.9:

Table 7.9 A 2 (ethnicity) $\times$ 3 (major) design with unequal n.

	Psychology	*Nursing*	*Hotel Administration*
Asian	n = 11	n = 11	n = 11
Hispanic	n = 11	n = 11	n = 10

There are equal sample sizes across the board except for Hispanics who are in Hotel Administration which is one short at 10. Rather than randomly discarding five pieces of data (one from the other five groups), Yates suggested that we add a cell mean to the group with the smaller sample size in order to make all groups even. The reasoning was that the cell mean would change neither the mean for that group (or overall) nor the variance. We would take the mean for the cell of the Hispanics who are in Hotel Administration and make an 11th subject, which is the cell mean. Then, one can perform the ANOVA with 11 subjects for all the groups. The original ANOVA summary table for the source and degrees of freedom is shown in Table 7.10.

Table 7.10 Source and df for the ethnicity × major design with the added cell mean.

Source	df
Ethnicity	1
Major	2
Ethnicity × Major	2
Within	60
Total	65

However, the problem is that you never ran that 11th Hispanic subject who was in Hotel Administration. Hence, you are not entitled to the extra power. Yates suggested that you deduct one degree of freedom from the within and total for each cell mean you add to attain equal n. The new summary table (source and df) is shown in Table 7.11.

Table 7.11 Final source and df for the ethnicity × major design using the Yates substitution formula

Source	df
Ethnicity	1
Major	2
Ethnicity × Major	2
Within	59
Total	64

Because of computers and more advanced approaches, this method is not really used today. The next two methods are more likely to be used by researchers, as they are in the standard statistical software packages.

3. Least Squares Solution

This is the most mathematically ideal of the procedures and is used by standard statistical software packages (e.g., SAS). In short, it is based on a multiple regression approach, which is extremely cumbersome to perform by hand. I will not pursue this procedure here, except to state that for 2 × 2 designs, the least squares solution and the next procedure, the unweighted means solution, are equivalent.

4. Unweighted means solution

In this approach, which is also used by standard statistical software packages (e.g., StatPac), the sum of squares effects (A, B, A × B) are calculated via the cell means, whereas the within or error term is calculated via the raw data. Suppose we have the following scenario:

Participants from Arizona and Nevada who designated themselves as either Protestants or Mormons gave a rating (from 0 to 9) as to whether there is life on other planets (0 represented no possibility at all; 9 represented absolutely there is life elsewhere). The hypothetical raw data for each group (totals are in parentheses; means are in brackets) is illustrated in Table 7.12.

Table 7.12 Hypothetical raw data for the state × religion unequal n design

	Protestants	*Mormons*	*Totals*
Arizona	5,3,4,7,8 (27) [5.4]	7,3,2,1 (13) [3.25]	(40) [8.65]
Nevada	2,4,6,1,6,8,3 (30) [4.2857]	5,7 (12) [6]	(42) [10.2857]
Total [Mean]	(57) [9.6857]	(25) [9.25]	(82) [18.9357]

We begin by calculating the sum of squares effects using the cell means.

SS state: $(\overline{T}_A^2 + \overline{T}_N^2)/n - (\overline{GT}^2)/N$

$(8.65^2 + 10.2857^2)/2 - (18.9357^2)/4 = .6688$

SS religion: $(\overline{T}_P^2 + \overline{T}_M^2)/n - (\overline{GT}^2)/N$

$(9.6857^2 + 9.25^2)/2 - (18.9357^2)/4 = .047458$

SS state × religion: SS total − SS state − SS religion

$4.449542 - .6688 - .047458 = 3.733284$

You will notice that this formula looks a little different from what you have seen before. That is, when we calculated the interaction, we used the sum of each cell squared and divided by n before subtracting out the correction factor and the sum of squares of each main effect. This really is the same formula; however, because we are dealing with means (rather than raw data), we are taking the square of each mean and dividing by 1 (number of means in that cell). This would be equivalent to $\sum\sum \overline{X}^2$ from the SS total.

SS total $= \Sigma\Sigma\overline{X}^2 - (\overline{GT})^2/N$

    $94.089724 - (18.9357^2)/4 = 4.449542$

In order to calculate the SS within, we will use the raw data.

    We start by calculating the sum of squares cells.

SS cells $= (T_{AC}^2/n_{AC} + T_{AM}^2/n_{AM} + T_{NC}^2/n_{NC} + T_{NM}^2/n_{NM}) - (GT^2)/N$

    $(27^2/5 + 13^2/4 + 30^2/7 + 12^2/2) - (82^2)/18 = 15.06587$

At this point, it might be more advantageous to compute the SS total (raw data), as we will need it for the calculation of the SS within.

SS within $= (SS \text{ total (raw data)} - SS \text{ cells}) / \overline{n}_h$

    $(92.44445 - 15.06587)/3.66 = 21.14$

$\overline{n}_h$ is called the **harmonic mean**. This is where you average the reciprocals and take the reciprocal of that average. In mathematical terms,

$\overline{n}_h$ = number of cells$/(1/n_{11} + 1/n_{12} + \cdots + 1/n_{jk})$

    $4/(1/5 + 1/4 + 1/7 + 1/2) = 4/1.092857 = 3.66$

In this formula, n_{11} represents the particular cell based on its respective row and column. That is, the first subscript "1" stands for the first row and the second subscript "1" stands for the first column. Therefore, in our example, the first subscript "1" stands for Arizona (the first row) and the second subscript "1" stands for Protestants (the first column). So, $1/n_{11}$, the sample size for Arizona Protestants would equal 1/5. Hence, at the end of the formula, j (which is a descriptor indicating the number of rows) would equal two and k (which indicates the number of columns) would also equal 2. Hence, if you saw $1/n_{22}$, then this would mean 1/the sample size for Nevada Mormons which would be 1/2.

The reason that we use the harmonic mean is because it weighs the cells with smaller sample sizes more and those with larger sample sizes a bit less. The basic idea is that if the experiment is run properly, then no cell should be weighted more than another cell, as they are all equally valid. If the arithmetic mean was used, then the cells with larger sample sizes would be weighted quite a bit more. Let's demonstrate this in a different scenario as shown in Table 7.13.

Table 7.13 Unequal n with the sample size for one cell 100 times larger than the others

	A1	*A2*	*A3*
B1	n = 5	n = 5	n = 5
B2	n = 5	n = 5	n = 500

The arithmetic average of the sample sizes of the six cells is 87.5. However, the average of the sample sizes using the harmonic mean is 5.98. As you can see, the n = 500 is weighted quite a bit more when using the arithmetic average, whereas it is weighted quite a bit less with the harmonic mean (hence, the term unweighted means).

SS total (raw data) $= \Sigma\Sigma X^2 - (GT^2)/N$

$466 - (82^2)/18 = 92.44445$

For the summary table, we could take the total 92.4445 and divide by the harmonic mean 3.66 for 25.258.

The summary table for the unweighted means solution is shown in Table 7.14.

Table 7.14 Summary table for the state $\times$ religion ANOVA with unequal n (in mean units)

Source	df	SS	MS	F	p
State	1	.6688	.6688	.443	p > .05
Religion	1	.047	.047	.031	p > .05
State × religion	1	3.733	3.733	2.472	p > .05
Within	14	21.141	1.51		
Total	17	25.258			

There were neither statistically significant effects for state or religion nor was there a statistically significant state $\times$ religion interaction. Remember that the effects were calculated with means, whereas the within was calculated with raw data. On the surface, these are two separate commodities like apples and oranges, so how are they comparable? If we think of the effects being calculated by means ($\Sigma X/N$) and the within term is calculated by raw data (ΣX), then if we divide the within term by N (or in this case, the harmonic mean), then the effects and the within are on the same scale (i.e., mean units). By the same token, instead of comparing mean units, perhaps one could compare the effects with the within term using raw data units. This is what the statistical software packages do. In order to do that, we would take each of the effects (state, religion, and state $\times$ religion) and multiply by the harmonic mean $(\Sigma\overline{X}/\overline{n}_h) \times \overline{n}_h$. Likewise, we could take the within term and not divide by the harmonic mean (SS total $-$ SS cells). The summary table would now be the following as shown in Table 7.15:

Table 7.15 Summary table for the state $\times$ religion ANOVA with unequal n (in raw data units)

Source	df	SS	MS	F	p
State	1	2.448	2.448	.443	p > .05
Religion	1	.174	.174	.031	p > .05
State × religion	1	13.664	13.664	2.472	p > .05
Within	14	77.379	5.527		
Total	17	92.444			

This is the summary table found in the statistical software packages that compute the unweighted means solution. Whether you deal with mean units or raw data units, the F ratios are the same. This is because multiplying or dividing by a constant will not change the value of F. These are called linear transformations. Values of F will change; however, if you use a nonlinear transformation such as a reciprocal, square root, or logarithm, just to name a few. Incidentally, if you had unequal n and a statistically significant main effect with more than two levels or a statistically significant interaction, then you could use range tests. The denominator of the q formula, however, would include the harmonic mean, rather than n.

The unweighted means solution may sound somewhat foreign to you. However, it is really an extension of the heuristic formula of F. The heuristic formula of F is $(ns_{\bar{x}}^2)/(\Sigma s_i^2/g)$. The numerator is composed of the variance of the means multiplied by n, or in this case, the harmonic mean $(\bar{n}_h)$. This keeps the numerator in raw data units (similar to how we multiplied each of the SS effects by the harmonic mean). Likewise, the denominator is in raw data units, given that it is the average of the within variances calculated from the raw data. We performed a similar task with regard to the within term for the unweighted means solution, as it was calculated from raw data.

All of these techniques are conservative and less powerful than having equal n. Nevertheless, given that unequal n occurs more often than not (primarily in survey work as opposed to straight experimental work), then it is important for us to have some rudimentary knowledge of these techniques. Our next section examines what happens when we obtain more than one score from each participant. These are called repeated measures designs.

Unequal n – Unweighted Means Solution – Class Example

A middle school teacher was interested in examining how well children can count money. She gave each student ten different amounts of money (including change) and the student had to tell her each time how much money was on the table. The total number of correct calculations by each student was the dependent variable. Here is the hypothetical data set.

	Male 7th Grade	Male 8th Grade	Female 7th Grade	Female 8th Grade
	4	4	2	8
	7	5	3	7
	4	9	1	9
		3	4	9
			1	7
				8
	_____	_____	_____	_____
Total	15	21	11	48
Mean	5	5.25	2.2	8

1. Show the ANOVA summary table.
2. Provide the appropriate conclusions.

CHAPTER 8

Repeated Measures Designs—Subjects × A

INTRODUCTION

Thus far, we have discussed independent groups or between subjects designs in which each subject provides only one score. However, there are designs in which we can obtain more than one score per subject. These are called repeated measures designs. For example, suppose you have three subjects and each one is tested three times on a particular measure (like the GRE Verbal, for instance). Here is how the data set would look as illustrated in Table 8.1.

Table 8.1 Hypothetical setup of a repeated measures design

	1	*2*	*3*
S1	152	153	157
S2	138	141	142
S3	144	146	148

There are a number of positive features to these types of designs. You will notice that there is only one score per cell, so there is no within-cell error term here. The between subjects designs have more than one score per cell (or per group); hence, you have a within term. Thus, there will be a different type of error term here. Moreover, there may be differences among subjects on their GRE Verbal scores. We can tease this out. Therefore, there will be a new term in the source table called "Subjects." As shown in Figure 8.1, the within or error term will be broken down into the subjects effect and the remainder will be error. Of course, the more subjects' variance one has, the lower the error term will be. This allows for greater power in these designs. Furthermore, because you are testing subjects repeatedly, there will be a more efficient use of subjects. Hence, in these designs, you will not need as many subjects in order to have reasonable power (e.g., .80). However, you don't get something for nothing in statistics. Although there are major advantages to repeated measures designs as we've just addressed (e.g., more efficient use of subjects, lower error term, greater power), there is an additional logical and statistical assumption (which will not be discussed in this book) that we need to address other than the standard statistical assumptions for ANOVA that we learned earlier. We begin by examining the Subjects × A or 1-within design.

Figure 8.1 Breakdown of the within term in a repeated measures design

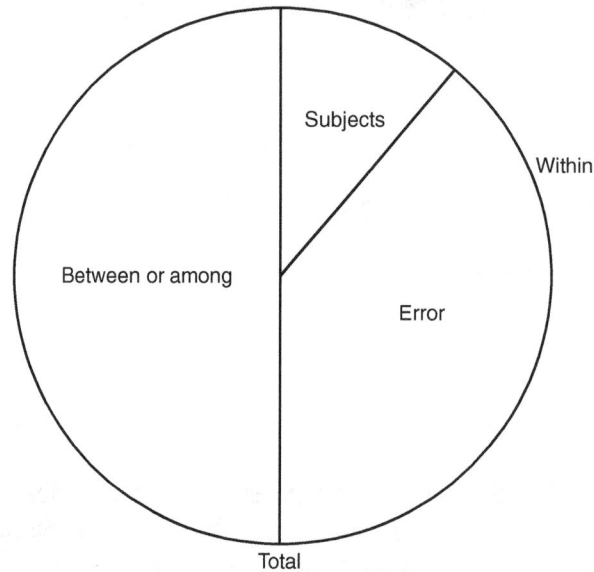

SUBJECTS X A

Suppose that a cognitive psychologist is interested in studying the vigilance and perceptual skills of the military. In this design, six soldiers were given a computer simulation task in which they had to shoot down 20 enemy targets in a 10-minute span. These targets were camouflaged. Each soldier was given three separate trials. The number of targets shot down was recorded. The data are shown in Table 8.2.

Table 8.2 Hypothetical data for the subjects × trials design

	T1	*T2*	*T3*	
Subject 1	5	8	9	= 22
Subject 2	3	5	4	= 12
Subject 3	11	13	17	= 41
Subject 4	9	11	12	= 32
Subject 5	7	8	14	= 29
Subject 6	2	7	12	= 21
	37	52	68	Grand Total = 157

What is the independent variable? Answer: Trials

What is the dependent variable? Answer: Number of enemy targets shot down

What is the null hypothesis in symbols? Answer: $\mu_{T1} = \mu_{T2} = \mu_{T3}$

What is the null hypothesis in words? Answer: There is no statistically significant difference among the population means of the three trials with regard to the number of enemy targets shot down.

SUMS OF SQUARES

We start with the sum of squares subjects, and it is exactly as you would expect.

SS subjects = $(T_{s1}^2 + T_{s2}^2 + T_{s3}^2 + T_{s4}^2 + T_{s5}^2 + T_{s6}^2)/n - (GT)^2/N$

   $(22^2 + 12^2 + 41^2 + 32^2 + 29^2 + 21^2)/3 - (157)^2/18 = 168.94$

We square the totals of each subject, add them, divide by n (the number of pieces of data that made up the total 22 for Subject 1 or 41 for Subject 2 or etc.), and subtract out the correction factor. For Subject 1, 5, 8, and 9 made up 22, so the n is 3. The grand total is 157, and N (the total number of pieces of data) is 18. This can be obtained by taking the number of subjects (6) times the number of pieces of data for each subject (3).

SS trials = $(T_{t1}^2 + T_{t2}^2 + T_{t3}^2)/n - (GT)^2/N$

   $(37^2 + 52^2 + 68^2)/6 - (157)^2/18 = 80.11$

Here we take the sum of each trial, square them, add them, and then divide by n (the number of pieces of data that made up 37, 52, or 68) which is 6. Again, we subtract out the correction factor.

SS subjects $\times$ trials = SS total – SS subjects – SS trials.

   $281.606 - 168.94 - 80.11 = 32.556$

Finally, we need to obtain the SS total.

SS total = $\Sigma X^2 - (GT)^2/N$

Take each piece of datum, square it $(5^2 + 8^2 + 9^2 + \cdots + 12^2)$, and add them, which equals 1651. Subtracting out the correction factor gives us 281.606.

SUMMARY TABLE

Now that we have the pieces to the puzzle, let's look at the summary table as shown in Table 8.3.

Table 8.3 Summary table for the subjects $\times$ trials design

Source	*df*	*SS*	*MS*	*F*	*p*
Subjects	5	168.94	33.78	(10.37)	(p < .01)
Trials	2	80.11	40.056	12.45	p < .01
Subjects $\times$ trials	10	32.556	3.256		
Total	17	281.606			

For the degrees of freedom of subjects, take the number of subjects and subtract one (ns – 1). Therefore, 6 – 1 = 5. Likewise, the same logic holds true for trials (number of trials – 1). Thus, 3 – 1 = 2. For the subject $\times$ trials interaction, take the degrees of freedom of subjects $\times$ degrees of freedom trials (5 $\times$ 2). The total degrees of freedom are N – 1 (18 – 1). Calculating the mean squares are still the same, namely, SS/df.

CONCLUSIONS

Obtaining the F ratio is a little different here. Unlike the independent groups designs, there is no within term. Hence, the subjects × trials interaction becomes the error term. To obtain the F for trials, take the MS trials/MS subjects × trials interaction. To determine statistical significance, we would go into the F table on 2 and 10 degrees of freedom. The critical values are 4.10 (05) and 7.55 (.01). The F of 12.45 is larger than both critical values, so the p < .01. The conclusion is that there is a statistically significant difference among the three trials with regard to the number of targets shot. Of course, one would need subsequent tests here in order to tell a more complete story. Regardless of subsequent test, the MS subjects × trials would be the error term (and the denominator degrees of freedom for the F or q tables).

You will notice that the F ratio for subjects is in parentheses along with its probability. The reason that these are in parentheses is because it is rare for this to be tested. In essence, the statistically significant Subjects effect indicates that the subjects differ from each other with regard to the number of targets shot. A statistically significant Subjects effect means that people differ, which is pretty much what psychology and the study of individual differences is all about. When you have 500 Introduction to Psychology students and you find a statistically significant Subjects effect, then it makes little to no sense to state that subject number 451 scored significantly higher than subject number 96. Therefore, when do we care about the Subjects effect?

The Subjects effect may be of interest when you have small sample sizes and know who the subjects are. For example, Subject 2 (Jamie) might be in the military for only a week as compared to Subject 4 (Pat) who had been in the military for three years. If there is a statistically significant difference in performance between these subjects, then perhaps part of that difference could be due to time spent in the military (i.e., training).

SUBJECTS × A INTERACTION

Now, let's have a look at the error term, the subjects × trials interaction. Suppose we have a hypothetical interaction as in Figure 8.2. Notice that all subjects are behaving consistently, thereby leading to a low interaction term (remember, a lack of parallelness in the simple effects). Hence, this lower interaction value, given that it is the error term, would yield greater power for the A (or trials) effect.

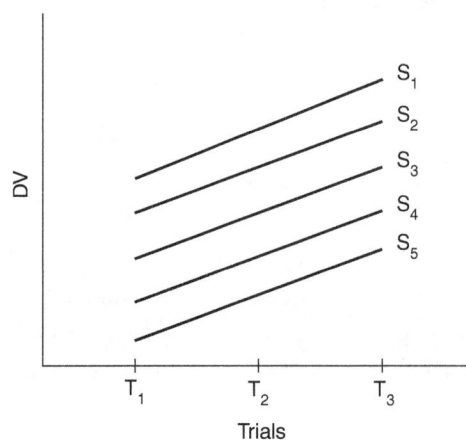

Figure 8.2 The subjects × trials interaction if subjects are consistent

However, in Figure 8.3, all subjects are not behaving consistently. There is a lack of parallelness in the simple effects here, thereby yielding a stronger interaction effect. Given that this is the error term, it would lower the power for the A (or trials) effect.

Figure 8.3 The subjects × trials interaction if subjects are not consistent

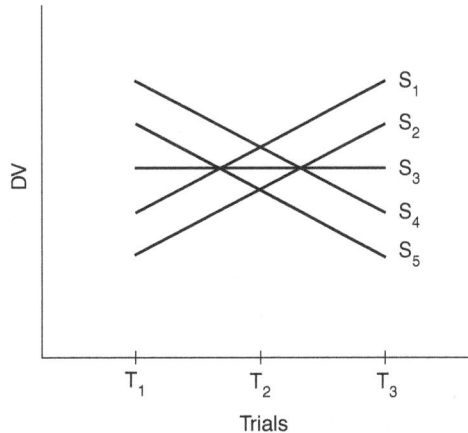

One of the problems of this design, in my humble opinion, is that this interaction has some interesting information that we cannot analyze. For example, it might be interesting to see how individual subjects trend across trials. Moreover, were there statistically significant differences across subjects over the last trial? Optimally, it would be wonderful if all subjects not only performed better over trials but we also examined the differences across subjects as a function of trial. For example, is there a statistically significant difference between Subject 5 (Dana, who had only one month of military experience but had excellent hand–eye coordination skills) and Subject 4 (Pat, who had more experience and was more consistent) at the end of the study? Because the interaction is the error term, there is no subsequent error term to test it. We now move to the logical assumption of repeated measures designs which is dealing with carry over effects.

Carryover Effects

The logical assumption of repeated measures ANOVA is to control for carryover effects. There are two types of carryover effects. The first is general carry over effects. These are general changes in the organism. Many of these are due to time, such as maturation, fatigue, boredom, practice, learning, and adaptation, just to name a few. If you are studying learning or maturation (especially if you are in developmental psychology), then carryover effect problems are not an issue. However, if we are in marketing or consumer psychology interested in determining if there is a difference among various types of cola products (e.g., RC Cola, Coke, Pepsi, and Big K Cola) in terms of taste, then by the time you get to the fourth cola, you may be tired of sampling cola. This could push the rating down, just because your taste buds are fatigued. Likewise, if you are a soldier in the military looking at a computer screen for an hour and shooting down enemy aircraft in a war simulation, then by the 30-minute mark, you might become progressively worse because of fatigue. Finally, if you are taking the GREs three times and the scores increase 3 points each time, might the minor increase each time be due to practice effects rather than actual learning?

The second type of carryover effect is specific carryover. Specific carryover is when the effect of treatment 1 followed by treatment 2 is different from the effect of treatment 2 followed by treatment 1. For example, if I shine a bright light in your eye followed by a dim light, the pupil will be different from if I shine a dim light first followed by a bright light. Moreover, if you run 220 meters followed by shooting five free throws, your heart rate would be different than if you shot five free throws first followed by running 220 meters.

So, how do you get rid of carryover effects? There are at least three ways. The first is through pre-experimental practice. If you have ever run an experiment (or any study) using subjects (even Introduction to Psychology students), then you will notice that after you read the instructions indicating what they should do in the task, there will be either a paucity or no questions. How often do you not ask questions in class for fear of having people think that you are stupid or you didn't listen properly? Although I am not much of a social psychologist, nevertheless, it might seem somewhat stigmatic to ask questions. Hence, subjects might state that they understand when in reality they might not totally. This can be a problem because subjects learn at different rates, so one subject who really understands might learn at Trial 1, whereas Subject 2 might learn at Trial 3, and Subject 3, who is somewhat confused, doesn't totally understand until Trial 7. Given these differences, the data are certainly problematic. Do you throw out the first eight trials? That is the impetus for providing pre-experimental practice when conducting an experiment. Let's provide an example of when it might be appropriate. Suppose you are providing a dichotic listening task. In this task, you would have different messages given in each ear. You would tell the subject that they need to shadow the left ear. For example, in the left ear, the message "a boy has a dollar and is buying Junior Mints," and in the right ear, "the horse is eating hay before running in the Kentucky Derby." You would then ask the subject what they heard in the left ear (both are somewhat simple sentences, so it should be easy for the subject to become familiar with the practice task). Problems with the instructions and answers can be rectified at this point. Once the subject is clear with regard to the instructions, then you can proceed to the main (and sometimes more difficult) task.

The second way of removing carryover effects is through the spacing out of trials. If we go back to our military example, perhaps we could give our soldiers a block of 10 minutes for shooting down enemy aircraft in a computer war simulation, followed by a 2-minute rest break. The soldier might be a bit more refreshed and fatigue would be neutralized. Hence, the difference between blocks might be a function of ability more than fatigue, which could be a potential confound in the study.

The final way of controlling for carryover effects is through randomizing the order of treatments. Random does not mean haphazard. There are actual mathematical algorithms for randomizing (see Brysbaert, 1991 for more details). For example, if all subjects performed tasks in the same order, then by the time they get to the final task, fatigue, boredom, or some other carryover effect may enter. Therefore, is the final task really harder (i.e., obtaining a lower score), or are the subjects doing poorly because of fatigue or boredom? By randomizing the order of the treatments, then the carryover effects are controlled.

Subjects × A (Trials)—Class Example

1-Within Design

A cognitive psychologist tested five subjects on a free recall task. The participants had to learn 10 words (e.g., sagacious, pusillanimous, and obsequious). Each word was presented every three seconds. There were five learning trials. Here are the results.

			Trials				
Subject	1	2	3	4	5		
1	7	8	8	9	10	=	42
2	2	4	6	7	8	=	27
3	3	5	4	5	7	=	24
4	5	4	6	5	9	=	29
5	6	6	6	7	8	=	33
Totals	23	27	30	33	42		(155)

1. What is the null hypothesis?

2. What are the independent and dependent variables?

3. Know how to obtain the summary table and the F ratio(s).

4. Draw appropriate conclusions.

Homework 6

Subjects × A (1-Within Design)

A human factors psychologist was interested in determining if there was a statistically significant difference in perceived hazard among signal words. The words were evaluated on a 10-point scale ranging from 0 = least hazardous to 9 = most hazardous.

Perform the ANOVA and complete the summary table. If there are statistically significant differences among the levels of the independent variable, then perform your choice of subsequent tests. Draw conclusions for the overall summary table and for the subsequent tests. You need not test the subject effects.

Subject	DANGER sign	WARNING sign	CAUTION sign
1	7	4	3
2	6	2	3
3	8	6	4
4	6	4	5
5	7	5	3
6	6	2	5
7	6	3	4
8	5	4	3

CHAPTER 9

Subjects × A × B design (2-within design)

This design is an extension of the Subjects × A (or 1-within design). Here we have two independent variables which are both repeated measures. For this example, we had five business travelers stay at the Hyatt, the Westin, and the Marriott for their business trips during the years 2015 and 2016. Each individual rated their stay on a scale from 1 to 10 with 1 being dreadful and 10 being excellent. Here are the data.

Table 9.1 Hypothetical satisfaction data for business travelers staying at the Hyatt, Westin, and Marriott in 2015 and 2016

	Hyatt		*Westin*		*Marriott*		
	2015	*2016*	*2015*	*2016*	*2015*	*2016*	
Subject 1	1	4	2	9	7	3	= 26
Subject 2	2	4	6	8	3	4	= 27
Subject 3	4	3	4	9	4	5	= 29
Subject 4	3	5	5	8	4	6	= 31
Subject 5	2	4	8	10	6	4	= 34
	12	20	25	44	24	22	Grand Total = 147

What are the independent variables? Answer: Hotels, Years

What is the dependent variable? Answer: Satisfaction ratings

What are the null hypotheses?

1. $\mu_H = \mu_W = \mu_M$
   There is no statistically significant difference between the population means of the hotels with regard to their satisfaction ratings.

2. $\mu_{2015} = \mu_{2016}$

There is no statistically significant difference between the population means of the years with regard to their satisfaction ratings.

3. There is no hotel $\times$ year interaction in the population.

SUMS OF SQUARES

If we think of this design as an extension of the Subjects $\times$ A design, then we can't go wrong. Therefore, similar to the former design, we start with the sum of squares subjects.

SS subjects $= (T_{s1}^2 + T_{s2}^2 + T_{s3}^2 + T_{s4}^2 + T_{s5}^2)/n - (GT)^2/N$

$(26^2 + 27^2 + 29^2 + 31^2 + 34^2)/6 - (147)^2/30 = 727.16 - 720.3 = 6.866$

Once again, we square the totals of each subject, add them, and then divide by n (the number of pieces of data that made up the total 26 for Subject 1 (1,4,2,9,7,3) 27 for Subject 2 (2,4,6,8,3,4) or etc.). The n, of course, is 6. The grand total is 147 and N (the total number of pieces of data) which is 30. Once again, this can be obtained by taking the number of subjects (5) times the number of pieces of data for each subject (6).

SS hotel $= (T_H^2 + T_W^2 + T_M^2)/n - (GT)^2/N$

$(32^2 + 69^2 + 46^2)/10 - (147)^2/30 = 69.80$

Take the sum of each hotel, square them, add them, and then divide by n (the number of pieces of data that made up 32, 69, or 46) which is 10. Next, subtract out the correction factor.

SS subjects $\times$ hotel — You will want to create Table 9.2 which contains the total satisfaction ratings for each subject as a function of hotel by adding over year.

Table 9.2 Total satisfaction ratings for the subjects $\times$ hotel interaction

	Hyatt	*Westin*	*Marriott*
Subject 1	5	11	10
Subject 2	6	14	7
Subject 3	7	13	9
Subject 4	8	13	10
Subject 5	6	18	10

To obtain Subject 1 for Hyatt, take Subject 1's Hyatt score at 2015 (1) + Subject 1's Hyatt score at 2016 (4). Do this for each subject and you will create the aforementioned table.

$(T_{S1H}^2 + T_{S1W}^2 + T_{S1M}^2 + T_{S2H}^2 + T_{S2W}^2 + T_{S2M}^2 + T_{S3H}^2 + T_{S3W}^2 + T_{S3M}^2 + T_{S4H}^2 + T_{S4W}^2 + T_{S4M}^2 + T_{S5H}^2 + T_{S5W}^2 + T_{S5M}^2)/n - (GT^2/N) - SS_{subjects} - SS_{hotel}$

$(5^2 + 11^2 + 10^2 + 6^2 + 14^2 + 7^2 + 7^2 + 13^2 + 9^2 + 8^2 + 13^2 + 10^2 + 6^2 + 18^2 + 10^2)/2 - (147^2/30) - 6.866 - 69.80 = 12.533$

Keep in mind that the n is obtained by asking how many pieces of data made up 5 (for Subject 1, Hyatt: 1 + 4), 11 (Subject 1, Westin: 2 + 9) or etc. Again, you could also take N (30) and divide it by the number of cells, groups, or levels of each variable combined (Subject 1, Hyatt; Subject 1, Westin; etc.). There are 15 different cells or groups, so 30/15 = 2.

This is similar to the Subjects × A format (Subjects, A, Subjects × A). Now, we move on to the repeated measure B and the interaction of Subjects × B.

SS year = $(T_{2015}^2 + T_{2016}^2) / n - (GT)^2/N$
$(61^2 + 86^2) / 15 - (147)^2 / 30 = 20.833$

Take the sum of each year and square them, add them, and then divide by n (the number of pieces of data that made up 61, 86) which is 15. Next, subtract out the correction factor.

SS subjects × year – You will want to make Table 9.3 which contains the total satisfaction ratings of each subject for a particular year by adding over hotel.

Table 9.3 Total satisfaction ratings for the subject × year interaction

	2015	*2016*
Subject 1	10	16
Subject 2	11	16
Subject 3	12	17
Subject 4	12	19
Subject 5	16	18

To obtain Subject 1 for 2015 (10), take Subject 1's 2015 score at Hyatt (1) + Subject 1's 2015 Westin score (2), and Subject 1's 2015 Marriott score (7). Do this for each subject and you will create the aforementioned table.

$(T_{S12015}^2 + T_{S12016}^2 + T_{S22015}^2 + T_{S22016}^2 + T_{S32015}^2 + T_{S32016}^2 + T_{S42015}^2 + T_{S42016}^2 + T_{S52015}^2 + T_{S52016}^2)/n - (GT^2/N) - SS_{subjects} - SS_{year}$

$(10^2 + 16^2 + 11^2 + 16^2 + 12^2 + 17^2 + 12^2 + 19^2 + 16^2 + 18^2)/3 - (147^2/30) - 6.866 - 20.833 = 2.333$

Once again, the n is obtained by asking how many pieces of data made up 10 (for Subject 1, 2015: 1 + 2 +7) or 16 (Subject 1, 2016: 4 + 9 +3) or etc. Again, you could also take N (30) and divide it by the number of cells, groups, or levels of each variable combined Subject 1, 2015; Subject 1, 2016; etc.). There are 10 different cells or groups, so 30/10 = 3.

SS hotel × year – For this interaction, we would add over subjects and create a table for each level of hotel and year as shown in Table 9.4.

Table 9.4 Total satisfaction ratings for the hotel × year interaction

Hyatt		*Westin*		*Marriott*	
2015	*2016*	*2015*	*2016*	*2015*	*2016*
12	20	25	44	24	22

$$(T_{H2015}^2 + T_{H2016}^2 + T_{W2015}^2 + T_{W2016}^2 + T_{M2015}^2 + T_{M2016}^2)/n - (GT^2/N) - SS_{hotel} - SS_{year}$$

$$(12^2 + 20^2 + 25^2 + 44^2 + 24^2 + 22^2)/5 - (147^2/30) - 69.80 - 20.833 = 22.067$$

The n is obtained by asking how many pieces of data made up 12 (add across all subjects for Hyatt 2015: 1,2,4,3,2), 20 (add across all subjects for Hyatt 2016: 4,4,3,5,4) or etc. You could also take N (30) and divide it by the number of cells, groups, or levels of each variable combined (Hyatt 2015; Hyatt 2016; etc.). There are six different cells or groups, so 30/ 6 = 5.

SS subjects × hotel × year: This can be obtained by taking the SS total and subtracting all of the sums of squares above. The sums of squares subtracted out are SS subjects, SS hotel, SS subjects × hotel, SS year, SS subjects × year, and SS hotel × year. Therefore, when calculating, you might want to do the SS total before obtaining the SS subjects × hotel × year. Incidentally, everything is subtracted out (all main effects and two-way interactions) because all of these terms are contained within this three-way interaction.

$$158.7 - 6.866 - 69.80 - 12.533 - 20.833 - 2.333 - 22.067 = 24.267$$

To obtain the SS total, it is the same formula as before.

SS total = $\Sigma\Sigma X^2 - (GT)^2/N$
   $879 - (147)^2/30 = 158.7$

Take each piece of datum (all 30 of them), square them, and add them ($1^2 + 4^2 + 2^2 + \cdots + 4^2$) which equals 879. Subtracting out the correction factor gives us 158.7.

SUMMARY TABLE

Let's take a look at the summary table as shown in Table 9.5.

Table 9.5   Summary table for the subjects × hotel x year design

Source	df	SS	MS	F	P
Subjects	4	6.866	1.71	(<1)	(p > .05)
Hotel	2	69.80	34.9	22.27	p < .01
Subjects × Hotel	8	12.533	1.56	(<1)	(p >.05)
Year	1	20.833	20.83	35.71	p < .01
Subjects × year	4	2.333	.583	(<1)	(p > .05)
Hotel × year	2	22.067	11.033	3.63	p > .05
Subjects × hotel × year	8	24.267	3.033		
Total	29	158.7			

Once again, the degrees of freedom for subjects, hotel, and year are all (the numbers of subjects, hotels, or years − 1). The degrees of freedom for every interaction are the products of the degrees of freedom for each of the effects used. For example, the subjects × hotel × year interaction is the degrees of freedom of subjects × degrees of freedom for hotel × degrees of freedom for year (4 × 2 × 1). Of course, you could also take the degrees of freedom for subjects × hotel (8) and multiply the degrees of freedom for year (1). The total degrees of freedom are N (total number of pieces of data) − 1.

Similar to the Subjects × A design, there will be no within as the error term. Therefore, what would be the appropriate error terms? If you know the two rules, then you cannot go wrong.

ERROR TERMS FOR SUBJECTS × A × B

First, for effects not involving subjects (i.e., hotel, year, hotel × year), use the Subjects × that effect interaction as the error term. For hotel, take the MS hotel (34.9) and divide by MS subjects × hotel (1.56) and the F will be 22.27. The degrees of freedom for obtaining statistical significance will be 2, 8 (df hotel and df subjects x hotel). For year, take the MS year (20.83) and divide by MS subjects × year (.583) and the F will be 35.71. The degrees of freedom for obtaining statistical significance will be 1, 4 (df year, df subjects x year). For the hotel × year interaction, take the MS hotel × year (11.033) and divide by MS subjects × hotel × year (3.033) and the F will be 3.63. The degrees of freedom for obtaining statistical significance will be 2, 8 (df hotel × year, df subjects × hotel × year).

Second, for effects involving subjects, use the largest-order interaction (the one with the most terms; subjects × hotel × year) as the error term. Normally, this is not performed for the vast majority of summary tables. Again, if you have a very small number of subjects and you have interest in them (as we discussed in the Subjects × A design), then there may be some utility in testing them. That is why the Fs and p values are in parentheses. Nevertheless, for the sake of completeness, let's go through each. For subjects, take the MS subjects (1.71) and divide by MS subjects × hotel × year (3.033) and the F will be less than 1.0. The degrees of freedom for obtaining statistical significance will be 4, 8 (df subjects, df subjects × hotel × year). For subjects × hotel, take the MS subjects ×

hotel (1.56) and divide by MS subjects × hotel × year (3.033) and the F will be less than 1.0. The degrees of freedom for obtaining statistical significance will be 8, 8 (df subjects × hotel, df subjects × hotel × year). For the subjects × year interaction, take the MS subjects × year (2.333) and divide by MS subjects × hotel × year (3.033) and the F will again be less than 1.0. The degrees of freedom for obtaining statistical significance will be 4, 8 (df hotel × year, df subjects × hotel × year). Let me also point out that the Fs being nonsignificant (and less than 1.0) are not always the case. There are times when effects involving subjects could be statistically significant.

CONCLUSIONS

Let's examine the conclusions. For hotel, we would conclude that there is a statistically significant difference among the hotels with regard to their satisfaction ratings. Because there are more than two levels, subsequent tests would be needed. These would include orthogonal comparisons, the Bonferroni test, the Scheffe test (although a bit conservative), and range tests. The error term for these comparisons is subjects × hotel.

For year, we would conclude that 2016 had significantly higher satisfaction ratings than did 2015. We know this by examining the means (for 2015, the mean would be $61/15 = 4.06$; for 2016, the mean would be 5.73). There is no statistically significant hotel × year interaction.

Subjects × A (day) × B (time) (2-Within Design) Class Example

A social psychologist examined how much TV was watched during a weekend. The number of hours that each subject spent watching is provided below. Here are the results:

Subject	Saturday AM	Saturday PM	Sunday AM	Sunday PM	
1	3	5	2	2	= 12
2	2	5	1	3	= 11
3	3	3	2	1	= 9
4	2	4	3	5	= 14
5	1	0	2	2	= 5
Totals	**11**	**17**	**10**	**13**	**51**

1. What are the null hypotheses?

2. What are the independent variables? What is the dependent variable?

3. Know how to obtain the summary table and the F ratio(s).

4. Draw appropriate conclusions.

Homework 7

2-Within Design – Subjects $\times$ A $\times$ B

In an attempt to measure the reaction time of peripheral vision, a researcher showed a red, green, blue, or yellow circle on a computer screen in which the display was a city street lined with buildings. The circle was positioned on the left (L), center (C), or right (R) of the computer screen. The reaction time was measured in seconds. In this case, the subject received all possible combinations of color and position.

Perform the ANOVA and complete the summary table. If there are any statistically significant differences for either independent variable or for the interaction, then perform subsequent tests and draw the appropriate conclusions. You need not test the subject effects.

The hypothetical data are provided down below.

| Subject | Red | | | Green | | | Blue | | | Yellow | | |
	L	C	R	L	C	R	L	C	R	L	C	R
1	1	2	3	4	1	3	2	3	4	2	3	4
2	7	7	6	5	6	7	8	9	7	8	6	7
3	5	6	4	4	5	6	4	3	6	4	5	6
4	1	2	1	2	3	2	2	3	4	3	2	1
5	8	4	6	4	7	6	8	7	6	8	7	6
6	6	8	6	7	8	6	7	6	6	7	6	9

CHAPTER 10

Subjects/A × B design (Mixed Design)

This design is a hybrid between an independent groups design and a repeated measures design. The A effect, or how the subjects are grouped, would be the independent or between groups portion, whereas the B effect is the repeated measure or within subjects effect. For experimental designs that we addressed earlier, namely, the 2-within design (Subjects × A × B) or a 2-between design (A × B), what was designated as A or B was irrelevant. However, such is not the case here. The specific type of independent variable (either between or within subjects) must match the A or B effects.

Suppose Steve Wynn is interested in determining which slot machine is paying out more over a one-day period between his two hotels, the Wynn and the Encore. The three slot machines are Lil Red, Giant's Gold, and Colossal Wizards. All these machines have similar characteristics (e.g., a main reel and a colossal reel). Eighteen different slot players (nine different people from each hotel) played all three machines for 160 minutes each. The results are provided in hundreds.

Table 10.1 Hypothetical data of slot machine payouts (in hundreds) for each subject at the Wynn and Encore

		Lil Red	*Giant's Gold*	*Colossal Wizards*		
	Subject 1	20	30	60	=	110
	Subject 2	30	50	80	=	160
	Subject 3	25	20	40	=	85
	Subject 4	40	30	50	=	120
Wynn	Subject 5	50	70	40	=	160
	Subject 6	30	40	70	=	140
	Subject 7	60	30	80	=	170
	Subject 8	50	40	60	=	150
	Subject 9	40	40	50	=	130
		345	350	530	Total =	1225

(continued)

**Table 10.1 Hypothetical data of slot machine payouts
(in hundreds) for each subject at the Wynn and Encore** *(continued)*

		Lil Red	*Giant's Gold*	*Colossal Wizards*		
	Subject 10	70	50	10	=	130
	Subject 11	50	20	50	=	120
	Subject 12	80	40	30	=	150
	Subject 13	90	50	20	=	160
Encore	Subject 14	60	30	20	=	110
	Subject 15	40	70	10	=	120
	Subject 16	60	40	60	=	160
	Subject 17	50	80	30	=	160
	Subject 18	80	90	40	=	210
		580	470	270	Total =	1320

What are the independent variables? Answer: Hotels, slot machines

What is the dependent variable? Answer: Amount paid

What are the null hypotheses?

1. $\mu_W = \mu_E$
   There is no statistically significant difference between the population means of the hotels with regard to the amount paid by the slot machines.

2. $\mu_L = \mu_G = \mu_C$
   There is no statistically significant difference among the population means of the three slot machines with regard to the amount paid.

3. There is no hotel $\times$ slot machine interaction in the population.

SUMS OF SQUARES

Because this is a hybrid design, we will begin by examining the between groups effect and its error term.

SS between subjects = $(T_{s1}^2 + T_{s2}^2 + T_{s3}^2 + T_{s4}^2 + T_{s5}^2 + T_{s6}^2 + T_{s7}^2 + T_{s8}^2 + T_{s9}^2 + T_{s10}^2 + T_{s11}^2$
$+ T_{s12}^2 + T_{s13}^2 + T_{s14}^2 + T_{s15}^2 + T_{s16}^2 + T_{s17}^2 + T_{s18}^2)/n - (GT)^2/N$

$(110^2 + 160^2 + 85^2 + 120^2 + 160^2 + 140^2 + 170^2 + 150^2 + 130^2 + 130^2 + 120^2 + 150^2 + 160^2 + 110^2 + 120^2 + 160^2 + 160^2 + 210^2)/3 - (2545)^2/54 = 4730.093$

The formula is similar to the subjects' effects of the previous designs. n is 3 because that is how many pieces of data it took to calculate 110, 160, 85, or any of the subjects totals. The N is 54 because each subject (18) gave 3 scores. Therefore, $18 \times 3 = 54$. Again, to obtain n, we could take N and divide by the number of subjects $54/18 = 3$.

SS hotel $= (T_w^2 + T_E^2)/n - (GT)^2/N$

$(1225^2 + 1320^2)/27 - (2545)^2/54 = 167.13$

In the SS hotel, we add up all the values from the Wynn and all the values from the Encore, square them, add them, and subtract out the correction factor. How many pieces of data did you add from the Wynn or the Encore? That represents n (or 27). Again, one could take N (54) and divide by the number of levels of the independent variable (2).

SS subjects/hotel: How are subjects divided by? They are divided by hotel. Subjects tested the three slot machines in either the Encore or the Wynn, but not both. This is the between subjects error term. That is why it is important to know the difference between the A and B effects in this design.

SS subjects/hotel = SS between subjects − SS hotel
   $4730.093 - 167.13 = 4562.963$

Let us now calculate the repeated measures portion of the design.

SS within subjects = SS total − SS between subjects

Normally, we would calculate this a bit later. Yet, if you think about how you obtained the SS within in a one-way ANOVA, it was the SS total − SS between or among. The same idea works here as well (as long as you are thinking about subjects).

$22080.093 - 4730.093 = 17350$

SS slot machine $= (T_L^2 + T_G^2 + T_C^2)/n - (GT)^2/N$

$(925^2 + 820^2 + 800^2)/18 - (2545)^2/54 = 500.926$

In order to obtain the totals for slot machine, add all 18 scores for each. How many scores made up the total for Lil Red, Giant's Gold, or Colossal Wizards? That is how you obtain n.

SS hotel × slot machine: $(T_{WL}^2 + T_{WG}^2 + T_{WC}^2 + T_{EL}^2 + T_{EG}^2 + T_{EC}^2)/n - (GT^2)/N - SS \text{ hotel} - SS \text{ slot}$
   machine

$(345^2 + 350^2 + 530^2 + 580^2 + 470^2 + 270^2)/9 - (2545)^2/54 - 167.13 - 500.926 = 7456.481$

How many pieces of data made up the total of Lil Red at Wynn, the total for Giant's Gold at Wynn, etc.? There were 9 pieces of data that made up each total. Again, you could take N (54) and divide by the number of cells (6) to obtain 9.

SS subjects/hotel × slot machine = SS within subjects − SS slot machine − SS hotel × slot machine

17,350 − 500.926 − 7456.481 = 9392.593

This will be the error term for the repeated measures effects. Once again, I would normally compute this after the SS within subjects was calculated.

SS total = $\Sigma\Sigma X^2/(GT)^2/N$

As we have done numerous times, square each of the 54 pieces of data and then subtract out the correction factor.

142,025 − (2545)²/ 54 = 22,080.093

Once again, it might be advantageous to calculate the SS total before computing the SS within subjects and SS subjects/hotel × slot machine terms.

SUMMARY TABLE

Let's take a look at the summary table as shown in Table 10.2.

Table 10.2   Summary table for the Subjects/Hotel x Slot Machine design

Source	*df*	*SS*	*MS*	*F*	*p*
[Between Subjects]	17	4,730.093			
Hotel	1	167.13	167.13	.586	p > .05
Subjects/Hotel	16	4,562.963	285.185		
[Within Subjects]	36	17,350.00			
Slot Machine	2	500.926	250.463	.853	p > .05
Hotel × Slot Machine	2	7,456.481	3,728.24	12.702	p < .01
Subjects/Hotel × Slot Machine	32	9,392.593	293.519		
Total	53	22,080.093			

In the source column, the terms "Between Subjects" and "Within Subjects" are in brackets. This indicates to the reader that what follows is either between subjects or repeated measures based. Neither of these terms are tested, they are merely delineations of the design.

For the degrees of freedom,

Between subjects are ns − 1 (18 − 1 = 17), df hotel + df subjects/hotel (1 +16 = 17), or df total − df within subjects (53 − 36 = 17).

Hotel is h − 1 (2 − 1 = 1).

Subjects/hotel is obtained by taking the between subjects df − hotel df (17 − 1 = 16). The other df formula for subjects/hotel is a (n − 1). That is, (2) × (9 − 1) = 16.

Within subjects df can be obtained in a number of ways: df total − df between subjects (53 − 17 = 36), df slot machine + df hotel × slot machine + df subjects/hotel × slot machine (2 + 2 + 32 = 36), or na (b − 1), which is (9 × 2) × (3 − 1) = 36.

Slot machine df is simply s − 1 (3 − 1 = 2).

Hotel × slot machine is df hotel × df slot machine (1 × 2 = 2).

The subjects/hotel × slot machine is obtained by taking the df subjects/hotel × df slot machine (16 × 2 = 32). The formula is a (n − 1) (b − 1), which is 2 × (9 − 1) × (3 − 1) or 32.

The total df is N − 1 or (54 − 1 = 53). It could also be obtained by adding the between subjects and within subjects dfs (17 + 36 = 53) and the df hotel + df subjects/hotel + df slot machine + df hotel × slot machine + df subjects/hotel × slot machine (1 + 16 + 2 + 2 + 32 = 53).

These formulas are analogous to how you could obtain the SS as well. For example, for between subjects, you could add the SS hotel and SS subjects/hotel. Moreover, you could subtract the SS within subjects from the SS total.

In order to calculate the F for hotel, divide the MS hotel by the MS subjects/hotel error term. In order to determine statistical significance, the degrees of freedom would be 1, 16 (df hotel and df subjects/hotel). The F for slot machine and hotel × slot machine are obtained by taking the MS for slot machine and hotel × slot machine, respectively, and divide by the MS subjects/hotel × slot machine error term. For determining statistical significance for slot machine, the degrees of freedom are 2, 32 (df slot machine, df subjects/hotel × slot machine). Likewise, for hotel × slot machine, the degrees of freedom are 2, 32 (df hotel × slot machine, df subjects/hotel × slot machine).

CONCLUSIONS

Examining the summary table, the conclusions would be the following:

There was no statistically significant difference between the hotels with regard to the slot machine payout. There was no statistically significant difference among the slot machines with regard to their payouts.

However, there was a statistically significant hotel × slot machine interaction. Of course, we would need to plot the interaction and perform tests of simple effects followed up by range tests or just range tests themselves. Let me share a problem concerning the tests of simple effects. If we examine the simple effects of Wynn @ slot machine and Encore @ slot machine, then they would include the interaction coupled with the effect of slot machine. The correct error term would be subjects/hotel × slot machine, as both are contained in the repeated measures portion of the design. However, for the simple effects of Hotel @ Giant's gold, Hotel @ Lil Red, and Hotel @ Colossal Wizards, we are confounding the interaction with the main effect of hotel. The interaction portion has an error term of subjects/hotel × slot machine, but the effect of hotel has an error term of subjects/hotel. So, what is the proper error term? Because subjects/hotel is contained within subjects/hotel × slot machine, you cannot add up the error terms because they are not independent. Therefore, a pooling of the error terms and the degrees of freedom suggested by Satterthwaite (1946) would be used. If error 1 is Subjects/A and error 2 is Subjects/A × B, then to obtain the pooled the df denominator, the following equation is used:

df denominator = [SS error1 + SS error2]^2 / [(SS error 1^2/df error 1) + (SS error2^2/df error 2)]

(4562.963 + 9392.593)^2 / [(4562.963^2/16) + (9392.593^2/32)] = 47.99

In order to obtain the MS error term, use the following equation:

MS error = (SS error 1 + SS error 2)/(df error 1 + df error 2)

(4562.963 + 9392.593)/ (16 +32) = 290.74

If we had three hotels (Wynn, Encore, and Palazzo) and if the simple effect was significant for Hotel @ Lil Red, for example, then we would state that there was a statistically significant difference in payout for Lil Red across the three hotels. From there, we could perform range tests for the payout means of the three hotels for Lil Red using the Satterthwaite MS error term as part of the q equation. Likewise, you would also use the df denominator from Satterthwaite for the q table to evaluate the statistical significance of the pairwise comparison.

Although using the Satterthwaite correction is good practice, there are other degrees of freedom adjustment methods as well. One such method is the Kenward-Roger (1997), which expands on the Satterthwaite (1941) approximation. It has equal or better Type I error rates than the Satterthwaite correction in repeated measures designs (Schaalje et al, 2002). However, this procedure is a bit complex and will not be addressed here.

Subjects/A (bug problem) × B (pesticide)

1-between 1-within design Class Example

A human factors psychologist was interested in evaluating how careful one would be when using different types of pesticides. There were two groups of subjects:

1. Those who reported serious bug problems and used pesticides constantly.

2. Those who had few bug problems, rarely using pesticides.

The scale went from 1 to 8 (1= not very careful to 8 = extremely careful).

The hypothetical data are provided below:

Subject	Raid	D-Con	Holiday	Black Flag		
		Serious Bug Problem Group				
1	5	3	4	3	=	15
2	6	2	4	2	=	14
3	2	1	1	1	=	5
4	6	4	2	3	=	15
5	7	5	4	5	=	21
Totals	26	15	15	14		70
		No Bug Problem Group				
6	8	4	7	5	=	24
7	6	3	1	4	=	14
8	4	3	4	3	=	14
9	5	3	5	5	=	18
10	6	5	6	5	=	22
Totals	29	18	23	22		92

1. What are the null hypotheses?

2. What are the independent variables? What is the dependent variable?

3. Know how to obtain the summary table and the F ratio(s).

4. Draw appropriate conclusions.

Homework 8

Mixed Design — 1-between, 1-within Subjects/A × B

A Nevada administrator was interested in determining if there was a difference between the University of Nevada-Reno (UNR) and the University of Nevada-Las Vegas (UNLV) students in psychology with regard to GPA (grade point average). An educational researcher tracked students through all four years at their respective institutions and the GPA was determined by year.

Complete the summary table and if there are any statistically significant effects for the interaction or any other independent variable (where applicable), then do your choice of subsequent tests. Draw conclusions and plot the interaction, if appropriate.

The hypothetical data set is provided below.

	Freshman	*Sophomore*	*Junior*	*Senior*
UNR	1.8	3.9	1.4	1.7
	2.5	3.7	2.8	1.3
	1.7	3.9	2.0	1.6
	1.5	3.8	2.9	2.0
	1.6	3.9	2.0	1.9
	1.6	3.7	2.7	2.7
	2.7	3.8	2.0	2.5
UNLV	1.2	2.6	3.7	3.0
	2.1	1.9	3.2	3.6
	2.0	2.8	3.9	3.1
	2.2	1.0	3.0	3.2
	2.3	2.8	3.9	3.1
	2.4	2.2	3.1	3.0
	2.4	2.4	3.7	3.9

Rules for Determining the Type of Design

1. Is there only 1 score/subject?
   a. Yes: independent groups design (one-, two-, or three-way)
   b. No: repeated measures (go on to question 2)

2. Does every subject participate in every group and every condition?
   a. Yes: Subjects × A or Subjects × A × B
   b. No: (i.e., there is a grouping factor) Subjects/A × B (mixed design)

Helpful Tips for Answering Problems

A. How many independent variables are there?

B. Draw a diagram of the design.

C. Read carefully!! Be cognizant of words like "each," "or," "and," "either."

Sample Problems

1. 20 subjects were asked to rate three kinds of yogurt for taste, three times a day (morning, afternoon, evening). Each subject was tested with each yogurt and at each time of day in random orders.

2. 90 subjects were asked to rate one of three kinds of yogurt for taste. Subjects were tested in the morning, afternoon, or evening. Equal numbers of subjects participated in the various conditions.

3. 60 subjects were asked to rate three kinds of yogurt for taste. 20 subjects were tested in the morning, 20 in the afternoon, and 20 during the evening. Each subject was tested with each yogurt in random order.

Experimental Designs

For each of the following experimental designs, write out the first two columns of the ANOVA summary table (source and df). Be sure to include numbers where possible and draw arrows linking the particular effect with the appropriate error term. Answers to the experimental design problems are shown on the following pages.

1. 140 black rats were tested in a MED Associates avoidance box. The rats were equally divided into five different shock intensity groups. Latency of avoidance response was measured for each rat on each of 10 trials.

2. 500 human subjects with migraine headaches from Dr. Silver's classes were administered 0, 1, 2, 3, or 4 cc's of ALD403, a new experimental headache drug. The dependent variable was latency to relief of symptoms.

3. A researcher wanted to know whether men and women reacted differently to marijuana. 60 males and 60 females received one of five different dosage levels. After smoking the appropriate dosage, the subject was tested on a driving simulation task.

4. A researcher examined if e-cigarettes produced adverse effects over time. In this study, all the subjects were given an e-cigarette and were tested every 30 minutes for the number of adverse effects (e.g., nausea, cough) over a four-hour period. A group of 20 males and a group of 20 females participated.

5. 35 subjects witnessed six different scenarios of a robbery. All scenarios were two-minutes long and were randomized. The dependent variable was the number of physical characteristics remembered by the subject for each scenario.

6. Several undergraduate psychology students decided to test the idea that grade point average depends upon student disposition toward their professor. They selected 300 students of approximately equal grade point average and divided them randomly into two groups. One group was instructed to be polite and smile at the professor, and the other was to be impolite and not smile at their professor. At the end of the semester, the grade point averages were obtained and analyzed.

7. 24 subjects listened to the National Anthem sung by four female vocalists: Beyonce, Taylor Swift, Christina Aguilera, and Yuna. Each subject heard two different renditions of the National Anthem sung by each singer. Each subject listened to the selections in a random order. The response measure was a 10-point rating (1 being horrible to 10 being excellent).

8. 270 subjects were divided into three main subgroups. One group was tested after 8 hours of wakefulness, one after 16 hours of wakefulness, and the third after 24 hours of wakefulness. Each subgroup was broken down into smaller groups defined by three types of instructions. The dependent variable was the number of tasks completed on the in-basket test.

9. Three makes of cars Honda, Porsche, and Lamborghini, were used in an acceleration test. There were 27 2015 and 27 2016 cars equally divided by make. An equal number of cars of each year were filled with Chevron, Chevron with ethanol, or ethanol alone. The dependent variable was time to reach 60 mph from a standing stop.

10. 50 subjects were given the Attitudes Toward Cheating Scale by Gardner and Melvin (1988) and were equally divided into those who opposed cheating on exams and those who thought cheating was appropriate. Each subject took seven quizzes in a course with no proctor (but a camera was installed in the classroom). The dependent variable was the number of times that the subject engaged in cheating behavior.

Answers

1. Subjects/A × B

Source	df
[Between Subjects]	139
Shock	4
Subjects/shock	135
[Within subjects]	1260
Trials	9
Shock × Trials	36
Subjects/shock × trials	1215
Total	1399

2. One-way ANOVA

Source	df
Dosage	4
Within	495
Total	499

3. Two-way ANOVA

Source	df
Sex	1
Dosage	4
Sex × dosage	4
Within	110
Total	119

4. Subjects/A × B

Source	df
[Between subjects]	39
Sex	1
Subjects/sex	38
[Within subjects]	280
Time	7
Sex × time	7
Subjects/sex × time	266
Total	319

5. Subjects × A

Source	df
Subjects	34
Robbery scenario	5
Subjects × robbery scenario	170
Total	209

6. One-way ANOVA

Source	df
Disposition	1
Within	298
Total	299

7. Subjects × A × B

Source	df
Subjects	23
Vocalist	3
Subjects × Vocalist	69
Rendition	1
Subjects × rendition	23
Vocalist × rendition	3
Subjects × vocalist × rendition	69
Total	191

8. Two-way ANOVA

Source	df
Sleep	2
Instructions	2
Sleep × instructions	4
Within	261
Total	269

9. Three-way ANOVA

Source	df
Make	2
Year	1
Fuel	2
Make × year	2
Make × fuel	4
Year × fuel	2
Make × year × fuel	4
Within	36
Total	53

10. Subjects/A × B

Source	df
[Between subjects]	49
Cheating attitude	1
Subjects/cheating attitude	48
[Within subjects]	300
Quizzes	6
Cheating attitude × quizzes	6
Subjects/cheating attitude × quizzes	288
Total	349

CHAPTER 11

Correlation

INTRODUCTION

In the previous statistical procedure, analysis of variance (ANOVA), we examined if there were statistically significant differences between the means. We now examine a different statistic, namely, correlation. Correlation asks a different question. Is there a linear relationship between X and Y? Therefore, a **correlation** is a linear relationship between two variables. For example, such possibilities might include is there a correlation between home runs and rbi? Is there a correlation between pulse rate and heart rate? Is there a correlation between number of hours texting and GPA? Is there a correlation between body mass index and having Type II diabetes? For a realtor, is there a correlation between number of hours worked and number of houses sold?

The statistic used to examine correlation is called r, which stands for the Pearson product–moment correlation coefficient. This was named after the famous statistician Karl Pearson. Moreover, the value of r ranges from +1.0 to −1.0. Anything correlated with itself will always be 1.0.

Suppose we examine the blood alcohol level and injury severity score (Tulloh & Collopy, 1994) for drivers who had accidents. If we provide a hypothetical plot (scatterplot or scatter diagram), as shown in Figure 11.1, in which all points are in a straight line, then the correlation would be +1.0.

Figure 11.1 Hypothetical plot of blood alcohol level and injury severity score r = +1.0

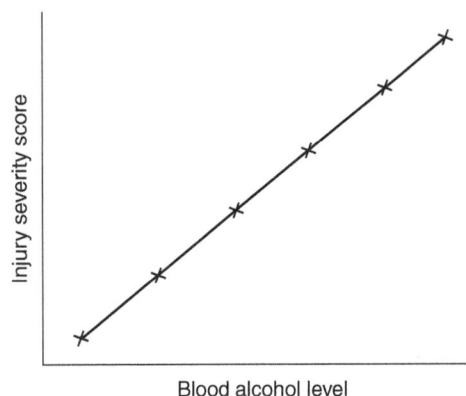

Injury severity score

Blood alcohol level

It is a positive correlation because the greater the X, the greater the Y. In other words, the lower the X, the lower the Y. This means that if you tell me what your X (blood alcohol level) is, then I can tell you perfectly what your Y (injury severity score) would be. In real research, the probability of having a perfect correlation is astronomical, as there is always error. Therefore, suppose the data have some error (how far from the line of best fit we are). As illustrated in Figure 11.2, the correlation still has a positive trend, but the correlation might drop to approximately +.50.

Figure 11.2 Hypothetical plot of blood alcohol level and injury severity score r = +.5

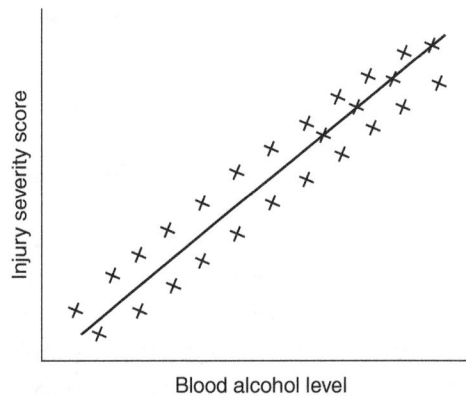

In a second example, if we correlate people's weights with SAT scores, then we would obtain a scatterplot similar to that found in Figure 11.3. If we examine the scatterplot, there are some folks who do not weigh much and have low SAT scores. There are some folks who do not weigh much and have moderate and high SAT scores as well. Likewise, those who weigh a lot have SAT scores all over the map. Therefore, there would be no linear relationship in this example or a correlation hovering around 0. It is never exactly absolute 0, because by chance alone, there will be some correlation (either slightly positive or slightly negative). This would indicate that there is no linear relationship between weight and SAT scores.

Figure 11.3 Hypothetical plot of weight with SAT scores r = 0

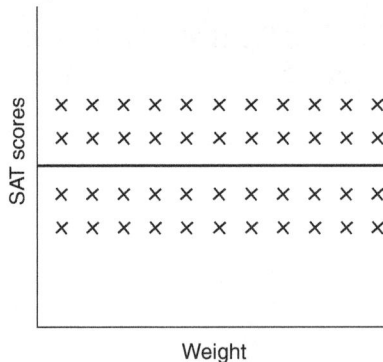

Our final example examines the correlation between temperature and precipitation rates in the summer (e.g., Madden & Williams, 1978). This would constitute a negative correlation or negative linear relationship. That is, as X increases, then Y declines; if X declines, then Y increases. In the hypothetical scatterplot, as shown in Figure 11.4, as the temperature increases, the amount of precipitation decreases. Conversely, as the temperature declines, then the amount of precipitation during these summer months increases. Hypothetically, the correlation would be roughly around −.40. Our next section will focus on the mathematical computation of r.

Figure 11.4 Hypothetical plot of temperature and precipitation rates during the summer months, r = −.4

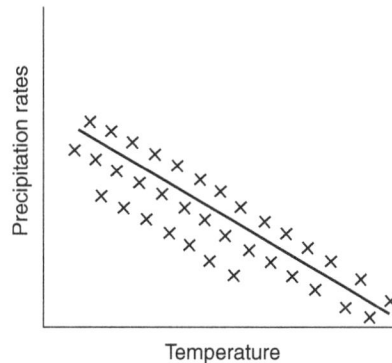

COMPUTATION OF r—SETTING THE SCENE

Let's demonstrate how to calculate the correlation both heuristically and computationally. Suppose 13 inmates from the Southern Desert Correctional Center in Indian Springs are administered the Clinical Anger Scale by Snell et al. (1995) and the Rosenberg Self-Esteem Scale (Rosenberg, 1965). The Clinical Anger Scale ranges from 0 to 63 in which 0–13 is no clinical anger, 14–19 is mild, 20–28 is moderate, and above 29 is severe clinical anger. The Rosenberg Self-Esteem Scale ranges from 0 to 30 in which 15–25 is normal and below 15 is low self-esteem. The results are shown in Table 11.1.

Table 11.1 Hypothetical results of anger and self-esteem scores of the 13 inmates

Anger	Self-Esteem
16	21
15	22
36	13
11	24
10	23
51	9
27	16
5	29
62	2
18	16
20	19
24	22
27	25

Inmate 1 scored 16 on anger and 21 on self-esteem, Inmate 2 scored 15 on anger and 22 on self-esteem, and so on.

HEURISTIC FORMULA OF r

In order to compute the correlation, we begin by examining the heuristic formula of r.

$$r = \frac{\sum z_x z_y}{N}$$

As we have seen with the other heuristic formulas, this may be a time-consuming procedure. We must first obtain the z-score for anger and the z-score for self-esteem for each subject. In order to compute the z-score, we need to calculate the population standard deviations (15.81 for anger, 7.012 for self-esteem) and the population means for each (24.769 for anger, 18.538 for self-esteem). For Subject 1, the z-score for anger is $(16 - 24.769)/15.81 = -.5546$; the z-score for self-esteem is $(21 - 18.538)/7.012 = .3510$. After the z-scores have been obtained for each subject, we then multiply each subject's z-score for anger and z-score for self-esteem in order to produce the product of their z-scores. For Subject 1, $-.5546$ (z-score for anger) $\times .3510$ (z-score for self-esteem) $= -.1946$. When we add the products of all the z-scores, we wind up with -11.8146. When divided by 13 (the number of subjects), then the correlation (r) is $-.909$. This information is shown in Table 11.2.

Table 11.2 Raw scores and z-scores for anger and self-esteem

Anger	Z_{anger}	*Self-Esteem*	$Z_{selfesteem}$	$Z_X Z_Y$
16	−.5546	21	.3510	−.1946
15	−.6178	22	.4936	−.3049
36	.7103	13	−.7858	−.5581
11	−.8703	24	.7789	−.6778
10	−.9340	23	.6362	−.5942
51	1.6589	9	−1.3603	−2.2566
27	.1411	16	−.3620	−.0510
5	−1.2503	29	+1.4919	−1.8653
62	2.3546	2	−2.3585	−5.5533
18	−.4281	16	−.3620	.1549
20	−.3016	19	.0658	−.0198
24	−.0486	22	.4936	−.0239
27	.1411	25	.9215	.1300
$\Sigma X = 322$		$\Sigma Y = 241$		$\Sigma z_x z_y = -11.8146$

$$r = \frac{-11.8146}{13} = -.909$$

Earlier in this chapter, we mentioned that the mathematical limits of correlation ranged from $+1$ to -1. Let's briefly examine how that occurs. First, you should remember that the mean of the z-scores is 0. Hence, when you sum up the z-scores for X or for Y, they will equal 0 ($\Sigma z = 0$). Yet, when you square each z_x or each z_y and sum them, you will obtain N ($\Sigma z^2 = N$; give it a try here, although there may be some minor rounding issues).

If z_x equals z_y, then in accordance with the formula, it would be equivalent to having $z_x \times z_x$ or z_x^2. If we sum each product, Σz_x^2 (which is N), and divide by N, then we would get $+1.0$. If z_x equals $-z_y$, then it would be equivalent to having $z_x \times -z_x$ or $-(z_x^2)$. If we sum each product, $\Sigma(-z_x^2)$ (which is $-N$) and divide by N, then we would get -1.0. If z_x does not equal z_y, then the correlation would range between $+1.0$ and -1.0.

COMPUTATIONAL FORMULA OF r

Given our past discussions with the heuristic formulas for variance and ANOVA, they are a bit more tedious and time consuming. Once again, it is time to provide the computational formula, which looks more formidable but is an easier approach. Here is the formula:

$$r = \frac{N\Sigma XY - \Sigma X \Sigma Y}{\sqrt{[N\Sigma X^2 - (\Sigma X)^2][N\Sigma Y^2 - (\Sigma Y)^2]}}$$

Examining the equation, you should know everything within it, except for ΣXY. This is $X \times Y$ for each of the 13 subjects and then summing those. The formula would be: $16 \times 21 + 15 \times 22 + 36 \times 13 + 11 \times 24 + 10 \times 23 + 51 \times 9 + 27 \times 16 + 5 \times 29 + 62 \times 2 + 18 \times 16 + 20 \times 19 + 24 \times 22 + 27 \times 25 = 4659$.

Placing everything into the formula, then we would obtain the following:

$$r = \frac{13(4659) - (322)(241)}{\sqrt{[13(11,226) - (322)^2][13(5107) - (241)^2]}} = \frac{-17,035}{18,738.482} = -.909$$

TESTING THE SIGNIFICANCE OF r

After obtaining a correlation of $-.909$, we now test the null hypothesis that $\rho = 0$. Specifically, that there is no correlation between anger scores and self-esteem scores in the population. Hence, ρ is the population correlation coefficient. If you like analogies, $\overline{X} : \mu$ as $r : \rho$ (sample mean is to the population mean as the sample correlation is to the population correlation). There are two tests that one could use, the F and the z-test. Let's first have a look at the F-test. Here is the formula:

$$F_{1,N-2} = \frac{r^2(N-2)}{1-r^2}$$

$$F_{1,11} = \frac{(-.909)^2(11)}{1-(-.909)^2} = 52.32$$

With the critical values of F being 4.84 (.05 level) and 9.64 (.01 level), it is quite clear that there is a statistically significant correlation (given that 52.32 is larger than both critical values). Our conclusion would be that there is a statistically significant negative correlation between anger and self-esteem scores such that the higher the anger score, the lower is the self-esteem score. Make sure that you use the term scores rather than actual anger or self-esteem as we never really examined the reality, just the scores. That is, don't reify.

Let's take a look at the equation for F, introduce the characters, and then explain how this formula is similar to an F formula that you have already seen (and hopefully, you know). First, the F has degrees of freedom of 1 and $N - 2$. The one is a constant because we have two variables minus 1 (similar to having 2 groups minus 1 in an ANOVA). The $N - 2$ can be thought of in two ways. How many points does it take to draw a straight line? I hope you said two. When you are drawing the line of best fit or regression line, there are two points that are fixed, and all other ones are free to vary (as long as they fall on the line). The second reason is that you lose a degree of freedom for each component calculated in a regression analysis, namely, the slope and the y intercept.

r^2 is called the **coefficient of determination**. It represents the proportion of Y (self-esteem score) variance accounted for by knowing X (anger score). For example, if all points fell on the line perfectly, then we would have a correlation of $+1.0$ or -1.0. This means 100% of the Y variance is accounted for by knowing X. Although we have not yet discussed regression or prediction, nevertheless, it would indicate that if you tell me what your anger score is, then I could tell you exactly what your self-esteem score would be. Obviously, that is a panacea. In our example, 82.6% of the Y (self-esteem scores) variance would be accounted for by knowing X (anger scores). If r^2 is the proportion of variance accounted for, then $1 - r^2$ is the proportion of Y variance not accounted for by knowing X (error variance). Hence, 17.4% of the Y (self-esteem scores) variance is not accounted for by knowing X (anger scores).

Does this F formula look like an old friend or an alien to you? Suppose we look at the heuristic formula of F for ANOVA: $(ns_{\bar{x}}^2)/(\Sigma s_i^2/g)$. On the surface, they appear to be different. But we can see the parallels. First, remember that $s_{\bar{x}}^2$ is the variance among the group means. You want this variance to be huge in order to increase power. This indicates the difference among the means. This is analogous to r^2. To the extent that r^2 is large (much different than 0), then the power will also increase. Moreover, this is the proportion of Y variance accounted for by knowing X. Thus, both pieces deal with variances. Likewise, $\Sigma s_i^2/g$, which is the average variance within groups and $1 - r^2$, the proportion of Y variance not accounted for by knowing X are both error variances. Therefore, the F ratio, regardless of whether you are dealing with means or correlations, are ratios of variances.

The second formula for testing the null hypothesis of $\rho = 0$ concerns the z-test. Before we address the z-test, let's briefly address the sampling distribution of r. The sampling distribution of r is illustrated in Figure 11.5. As you can see, when $r = 0$, then the distribution is normal. If variables are not correlated, then they are independent, or normally distributed. However, as the correlation becomes more positive or negative between variables, then you will obtain more skew. If the correlation is negative, then you obtain greater positive skew in the distribution, whereas if the correlation is positive, then the distribution has more negative skew.

Figure 11.5 Sampling distribution of r

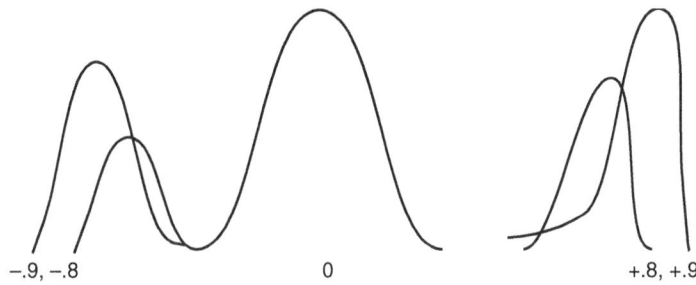

−.9, −.8   0   +.8, +.9

Because the sampling distribution of r is non-normal for correlations other than 0, Fisher (1921) developed a transformation that allowed for the testing of $\rho = 0$ using a normal curve test (z) by removing most of the skew in the distribution of r. The transformation, which is an intermediate step, is called z′ (z-prime). Fisher's z′ transformation is ½ $\log_e ((1 + r)/(1 − r))$. This is the formula that is found in most textbooks. In reality, there are more terms to this formula (if you know Taylor-series expansion), but for all intents and purposes, the other terms are negligible. Nevertheless, there are two characteristics that you should know concerning Fisher's z′ transformation. The first is that the distribution is approximately normal. The second is that the standard error of Fisher's z′ is $1/\sqrt{N − 3}$.

Therefore, the formula is as follows:

$$Z = (z'_r − z'_\rho)/ (1/\sqrt{N − 3}).$$

We need to obtain the z′ for r. Going to the z′ table, you will notice that the z′ for −.909 is −1.52. In this case, you could use −.90 (not −.91 as you are not entitled to that little extra power). The z′ for ρ (which is 0) is indeed 0. Therefore, this part of the formula is dropped. To make the formula a bit cleaner via algebra, we can now change it to the following:

$$Z = z'_r \sqrt{N − 3}$$

$$Z = −1.522 \sqrt{10} = −4.812$$

Is the z-score larger than the magic numbers 1.96 (.05) or 2.58 (.01)? Obviously, the z-score is larger than the critical values. Regardless of whether the z-score is positive or negative, remember that it is two tailed, so the most important issue is how different is the correlation from 0? The conclusion would be the same here as it was for the F-test, namely, that there is a statistically significant negative correlation between anger and self-esteem scores such that the higher the anger score, the lower is the self-esteem score.

Whether you use an F-test or a z-test, there will be no substantive difference in the conclusions. In fact, when I was a graduate student (many moons ago), I performed a brief simulation and found the difference in the probabilities to be out in the 6th decimal place.

CONFIDENCE INTERVAL FOR r

When we compute a correlation, it is not a perfect measurement. In other words, we could compute the correlation again with a different set of data and wind up with a different value. It is hoped that the values would be relatively close, but there is no guarantee. The same, of course, holds true for the mean. We are always striving to obtain our best estimate of the parameter, whether it be μ, or in this case, ρ. Therefore, it is reasonable practice to provide a confidence interval for the statistic (r), in order to determine not only the interval which will fall around the parameter (ρ) but also if another researcher is performing a similar study, then they can compare their correlation to the values contained within your confidence interval to determine if there is an overlap. If there is none, then maybe there is something about their sample (or yours) that is a bit different (e.g., Harvard students, professional athletes, fanatics, people who make over 10 million dollars a year).

In order to get started, we take our correlation $r = -.909$ and convert that to z'. Go into the z' table and read from left to right, $r = -.909$; The z' is -1.522. Because the z' distribution is approximately normal, add a negative sign to the z'.

Second, if we are calculating the 95% confidence interval for r, then use the following equation:

$$z' \pm 1.96 \times 1/\sqrt{N - 3}$$

Applying this to our example, the equation is as follows:

$$-1.522 \pm 1.96 \times 1/\sqrt{10}$$

This will provide us with a $z'_{upper} = -2.14$ and a z'_{lower}, which is $-.902$. However, this is not a confidence interval for z'. This is a confidence interval for r. Therefore, we must go back to the z' table and now backtransform from z' to r (reading right to left in the table). The formula is $r = (e^{2z} - 1)/(e^{2z} + 1)$. The $r_{upper} = -.972$ and the $r_{lower} = -.717$. We would write the confidence interval for r in a similar manner as we did for the confidence interval of the mean.

$$\text{Prob } [-.972 < \rho < -.717] \ .95$$

We would state that there is a 95% probability that the interval from $-.972$ to $-.717$ will fall around ρ. I changed the upper and lower to account for the negativity.

If we are calculating the 99% confidence interval for r, then the equation changes slightly:

$$z' \pm 2.576 \times 1/\sqrt{N - 3}$$

Again, calculating this from our data, the $z'_{upper} = -2.336$ and a z'_{lower}, which is $-.707$. By backtransforming, the $r_{upper} = -.981$ and the $r_{lower} = -.608$. Analogous to our earlier example, we write it out in the following way:

$$\text{Prob } [-.981 < \rho < -.608] \ .99$$

There is a 99% probability that the interval from −.981 to −.608 will fall around ρ.

Let's recap the steps for obtaining a confidence interval for r.

1. Convert r to z′

2. Use the equation: $z' \pm 1.96$ (or 2.58 for the 99%) $\times 1/\sqrt{N-3}$

3. Calculate z'_{upper} and z'_{lower}

4. Backtransform from z'_{upper} to r_{upper} and from z'_{lower} to r_{lower}

5. Place in the following format: Prob [$r_{lower} < \rho < r_{upper}$] .95 or .99

Did you notice that this formula is similar to the formula for the confidence interval of the mean? Let's take another look at the confidence interval formula of the mean.

$$\overline{X} \pm t_{df} \, s/\sqrt{N}$$

You will note that this is the sample mean multiplied by the t value on its degrees of freedom. If the N is above 100, then (for all intents and purposes) the t_{df} will change to z (i.e., the magic numbers 1.96 or 2.58). Finally, $s/\sqrt{N}$ is the standard error of the mean.

$$z' \pm 1.96 \times 1/\sqrt{N-3}$$

Analogously, z′ is multiplied by the z-score of 1.96 or 2.58 (depending on whether you want a 95% or 99% confidence interval). Moreover, $1/\sqrt{N-3}$ represents the standard error of z′. Therefore, there are distinct parallels between the two formulas.

OBTAINING A GOOD ESTIMATE OF ρ

There are many factors that affect the size of the correlation coefficient. However, one with which we are familiar is sample size. As you remember, the sample mean ($\overline{X}$) was the best estimate of the population mean (μ). The estimate of the population mean was more accurate as the sample size increased. When the sample size is small, the correlation coefficient is a highly variable statistic. The values for r can be all over the map. When sample size is increased, then we obtain a more stable estimate of ρ. However, this brings up an interesting issue. Suppose that you are testing a select group in which you have a small sample size (e.g., commercial pilots, hotel housekeepers, children with Asperger's syndrome, pianists, or pigeons). How might you obtain the best estimate of ρ? You could test these participants repeatedly (e.g., four times). This way you could compute correlations between Times 1 and 2, Times 1 and 3, Times 1 and 4, Times 2 and 3, Times 2 and 4, and Times 3 and 4. Which correlation of the six would be best? Obviously, they are all valid, so one could arithmetically average the correlations to obtain the best estimate of ρ. Another possibility would be to take the correlations, convert them to Fisher z's, average the z's, and then backtransform to r. Suppose the correlations are .30, .40, .70, .30, .60, and .70. The arithmetic average would be .50. Performing the backtransformed average z, we would add the z's for each correlation (.3095 + .4236 + .8673 + .3095 + .6931 + .8673 = 3.4703/6 = .5783), backtransforming to r, we would obtain an r = .521. Which estimate is better? Silver and Dunlap (1987) showed that the backtransform average z provided a better estimate of the population correlation coefficient and had a smaller standard error when the correlation was above .50. When the correlation was below .50, the waters were a little muddier. The backtransform average z was a better estimate of ρ, but had a larger standard error than arithmetic average r. Nevertheless, they recommended using the backtransform average z.

Correlation and Causality

In general, it is extremely difficult to infer causality from a correlation. There are many textbooks stating that you can only discuss relationships. For all practical purposes, this is true. However, there is a way to infer causality from correlation, provided the following two statements are satisfied.

1. If X causes Y, then Y cannot cause X. For example, if excessive alcohol consumption causes vomiting, then vomiting cannot cause excessive alcohol consumption. That one makes some sense, as vomiting does not cause excessive alcohol consumption. In fact, when one is vomiting due to this, perhaps tomato juice would be a better cure.

2. If X causes Y, then no third variable can cause either X or Y. This means if excessive alcohol consumption causes vomiting, then no third variable like stress can cause either excessive alcohol consumption or vomiting. This one is extremely difficult to prove. Obviously, stress can cause both. In my case, watching my beloved Cincinnati Reds, Cincinnati Bengals, or Sacramento Kings lose could cause both. Therefore, it is extremely difficult to eliminate all possible third variables. Hence, this is why we discuss relationships when using correlations as opposed to causality. Although it is hypothetically feasible, it is easier said than done. If we were discussing causality, then we would need to randomly assign subjects to groups, control all extraneous variables, and then determine if there are differences in the dependent variable across groups.

Assumptions of correlation

Similar to ANOVA, there are assumptions of correlation. Do you remember the assumptions from ANOVA? If you do, then these are reasonably analogous.

1. **Normality in the Arrays**. First, we should define what an array is. An **array** is all Y values for a given X. For example, suppose that I had 10,000 inmates and 100 of them had an anger score of 10. If I plotted the frequency distribution of all 100 self-esteem scores for those inmates, then I would obtain a distribution similar to the following found in Figure 11.6.

Figure 11.6 Normality in the arrays

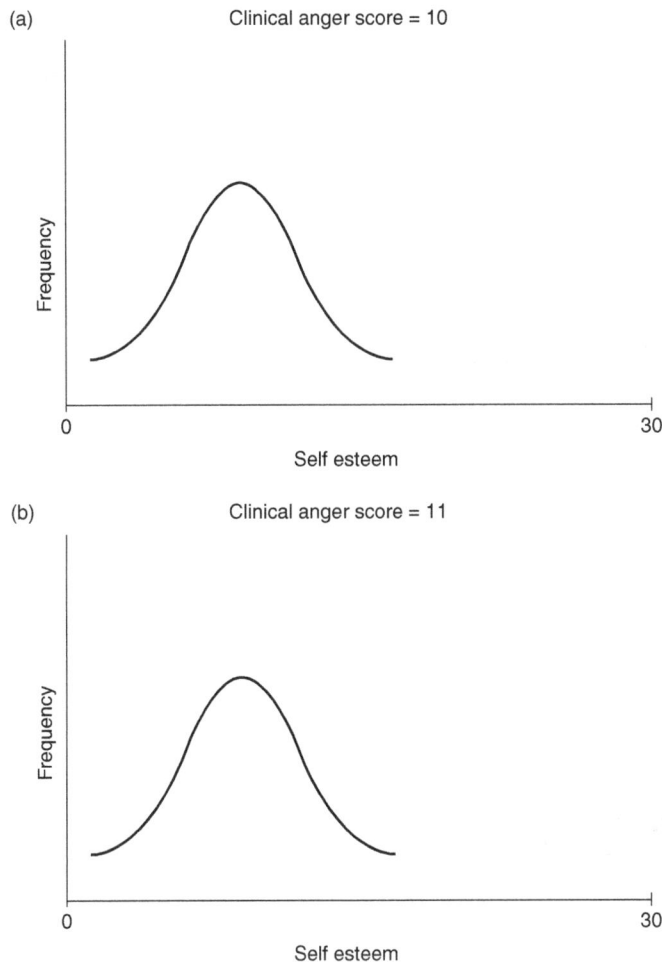

(a)   Clinical anger score = 10

(b)   Clinical anger score = 11

You will notice that the distribution is normal. If I examined the distribution of all self-esteem scores for all inmates with anger scores of 11, then I would again obtain a distribution that was normal. This would hold true for all possible arrays.

2. **Homogeneity of Variance in the Arrays**. Once again, suppose that I had 10,000 inmates and 100 of them had an anger score of 10. If I computed the variance of the 100 self-esteem scores for those who had an anger score of 10, then that variance should be essentially equal to those inmates self-esteem scores who had an anger score of 11, and so on.

3. **Linearity**. If we are dealing with correlation, which is a linear relationship between two variables, then it only makes sense that linearity would be an assumption.

   If you violate these assumptions when testing $\rho = 0$, then we will assume that the statistic is still robust (Edgell & Noon, 1984; Havlicek & Peterson, 1977).

   Note that the first two assumptions are analogous to the ANOVA assumptions of normality in the population and homogeneity of variance in the population. Moreover, we also stated that the F-test was robust to violation of the assumptions (although there were additional caveats mentioned).

Restriction of Range

Suppose we are interested in examining the correlation between children's age and psychomotor ability. Psychomotor ability deals with movement, such as how long one can keep their wand going around a circle in a rotary pursuit task. As age increases from 5 to 12, psychomotor ability also increases. This would make sense because as we age during childhood, then the skills become more refined and habitual. Therefore, the correlation would be a positive one, roughly around .40. However, if we are interested in examining young children (5 and 6 year olds) and obtain a snapshot of the data, then it would look like the following as illustrated in Figure 11.7.

Figure 11.7 Hypothetical scatter plot of age and psychomotor ability with restriction of range

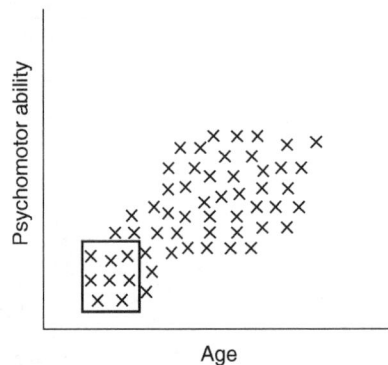

The correlation between age and psychomotor ability would hover around 0. Therefore, when we restrict the range in a linear relationship, then we will underestimate what the true population (ρ) will be.

In a different scenario, we examine the correlation of temperature with performance on a statistics test. If the temperature is similar to that of Antarctica, then you will probably perform poorly. You will be concentrating more on obtaining heat rather than focusing on the test. By the same token, if you take the statistics test in the blazing heat of the Las Vegas sun in the summer (about 115 happy degrees), then that will also affect cognitive processing in a negative way. Of course, if we provide a temperature ranging from 68-75 degrees, then it is hoped that the comfort of this temperature will optimize performance (provided you studied, of course). Therefore, the correlation between temperature and performance is approximately 0. This is because there is a curvilinear relationship rather than a linear one. This is why it is important to state that there is no correlation or no linear relationship rather than stating that there is no relationship. There is certainly a relationship here (i.e., curvilinear), just not a linear one. Nevertheless, if we restrict the range of temperature between 75 and 120 degrees, then we would wind up with a higher negative correlation than what we would expect (see Figure 11.8). Likewise, if we restrict the range of temperature between −40 and 68 degrees, then we would have a higher positive correlation than what we would expect. In short, when we restrict the range in a curvilinear relationship, then we will overestimate what the true population correlation (ρ) will be.

Figure 11.8 Hypothetical plot of temperature and statistics test performance with restriction of range

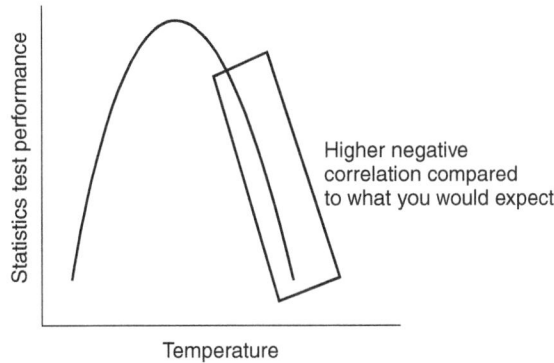

Hypothesis Testing: ρ = other than 0

The only null hypothesis test that we have tackled thus far is $\rho = 0$. In a slightly different problem, suppose you are interested in the level of organizational (university) commitment and GPA. When examining the literature, you find that Study A conducted at the University of Oregon had a correlation of .25; Study B conducted at UCLA had a correlation of .15; and Study C conducted at UNR had a correlation of .20. You conducted the study at UNLV and obtained a correlation of .40 for 100 undergraduate students. The question here is as follows: Is the correlation that you computed (.40) significantly higher than what is out in the literature $\rho = .20$ (the average of the three correlations)? The same z formula that we discussed concerning the null hypothesis of $\rho = 0$ would be applicable here: $Z = (z'_r - z'_\rho)/(1/\sqrt{N-3})$. In this case, z'_ρ is now .20 rather than 0, so we need to keep it in the equation. If we perform a little algebra, then we can modify the equation to the following: $Z = (z'_r - z'_\rho) \times \sqrt{N-3}$. Applying our information, $Z = (.424 - .203) \times \sqrt{97} = 2.176$. Because this z-score is larger than the critical value of 1.96 (.05 level), then $p < .05$. We would conclude that UNLV had a significantly higher correlation between university commitment and GPA than the average of the correlations set forth in the literature.

Testing the Difference Between Independent Correlations

Suppose the Nevada Board of Education was interested in determining if there was a difference in the correlation between total number of hours texting during the academic year and total score of the end of the year exams between Bishop Gorman and Bonanza high schools. For Bishop Gorman, the correlation was −.25 and 60 students were sampled. For Bonanza, the correlation was −.41 and 200 students were sampled. The reason that this is a test of two **independent correlations** is because we are dealing with two separate groups (Bishop Gorman and Bonanza students). The null hypothesis would be $\rho_1 = \rho_2$. In words, there is no statistically significant difference between the population correlations of total number of hours texting during the academic year and total score of the end of the year exams for Bishop Gorman and Bonanza high schools.

In order to determine if there was a difference, the following formula is used:

$$Z = \left(z_1' - z_2'\right) \Big/ \left(\left(1 / \sqrt{N_1 - 3}\right) + \left(1 / \sqrt{N_2 - 3}\right)\right)$$

$$Z = \left(-.255 - (-.436)\right) \Big/ \left(\left(1 / \sqrt{57}\right) + \left(1 / \sqrt{197}\right)\right) = 1.19$$

Because the Z was less than 1.96 or 2.58, then we would fail to reject the null hypothesis, $p > .05$. Hence, there was no statistically significant difference between the correlations of total number of hours texting during the academic year and total score of the end of the year exams for Bishop Gorman and Bonanza high schools.

If the Nevada Board of Education wanted to include Cimarron and Centennial high schools as well, then the null hypothesis would be $\rho_1 = \rho_2 = \rho_3 = \rho_4$. The problem here is that the Z-test addresses only two groups at a time. Therefore, if we use this test, then we would need to perform it six times examining all pairs of correlations. Of course, this leads to a compound of Type I error rate, which means that the more tests that we perform, by chance alone, we will find a difference that will be statistically significant. There are other procedures for testing this hypothesis and keeping Type I error rate within some control. In essence, this would be similar to a one-way ANOVA with four groups. Next, it would be followed up with subsequent range tests. However, this topic is beyond our undergraduate course.

On a final note, if you can test the difference between independent correlations, then you can also test the difference between dependent correlations. **Dependent correlations** occur within the same sample. For example, suppose you wanted to test the hypothesis that the correlation between job satisfaction and salary is higher than the correlation between number of years of experience and salary. This would be the test of $\rho_{12} = \rho_{13}$, that is the test of two dependent correlations with one element in common. The element in common is salary. In a different scenario, if you wanted to test the hypothesis that the difference between the correlation of job satisfaction and salary at Time 1 and job satisfaction and salary at Time 2 (after a 6% increase for everyone as a December bonus), then this would be the test of $\rho_{12} = \rho_{34}$. In other words, this is the test of two dependent correlations with no elements in common. Although you might believe that they are common (job satisfaction and salary), the fact that they are measured at different times means that they are separate entities. In other words, you would have scores for job satisfaction at Time 1 and Time 2 and scores for salary at Time 1 and Time 2. Therefore, the correlation matrix of these four variables would yield six correlations (all pairwise variables, similar to the matrix for range tests) that would be used in computations. The standard significance tests formulas for testing these hypotheses are somewhat complicated and beyond our course.

Linear Regression

One of the most important facets of statistics (in my humble estimation) is prediction. The term regression, means prediction. In **linear regression**, this is where you predict Y from X. In our scenario, we will be predicting self-esteem scores from anger scores. In order to perform this procedure, we will be drawing the line of best fit (regression line). Do you remember the formula for a line? If you said $y = mx + b$, then that is what we learned when we were children. Leave it to statisticians to complicate matters. The formula for linear regression is $\hat{Y} = bX + A$. $\hat{Y}$ is the predicted Y, b represents the slope (the tilt of the line), and A is the Y intercept (where on the Y-axis should the line intercept or cross, if we stretch it out that far). Let's use our example to demonstrate the procedure. First, we need to calculate the slope. The formula for slope is $b = r \times (s_y/s_x)$

This is the correlation multiplied by the ratio of the sample standard deviations.

$$-.909 \times (16.45/7.298) = -.403$$

Obviously, if the correlation is negative, then the slope will be negative as well.

To obtain the Y intercept, the formula is

$$A = \overline{Y} - b\overline{X}$$

$$18.538 - (-.403)(24.769) = 28.52$$

Therefore, the linear regression formula is as follows:

$$\hat{Y} = bX + A$$

$$\hat{Y} = -.403X + 28.52$$

In order to obtain the regression line, use any value of X (preferably one that would encompass a value on the scatterplot) to predict the Y value as illustrated in Table 11.3.

Table 11.3 Raw score and predicted Y values for the anger and self-esteem example

X	$\hat{Y}$	
10	24.49	$(-.403\,(10) + 28.52)$
20	20.46	$(-.403\,(20) + 28.52)$
30	16.43	$(-.403\,(30) + 28.52)$

Figure 11.9 represents the **scatterplot** (the coordinates for each subject's anger and self-esteem scores) and the line of best fit (regression line). This regression line is our best estimate of what subjects' self-esteem scores would be for their given anger scores.

Suppose that the correlation is 0 (e.g., the correlation between anger scores and the last 4 numbers of their social security numbers). The slope, or $b = r \times (s_y/s_x)$, would be 0, because the r is 0 (and 0 times anything is 0).

The Y intercept: $A = \overline{Y} - b\overline{X}$, would be the mean of Y because the slope is 0 (therefore, the slope times the mean of X is 0).

Figure 11.9 Scatter plot and regression line of anger and self esteem

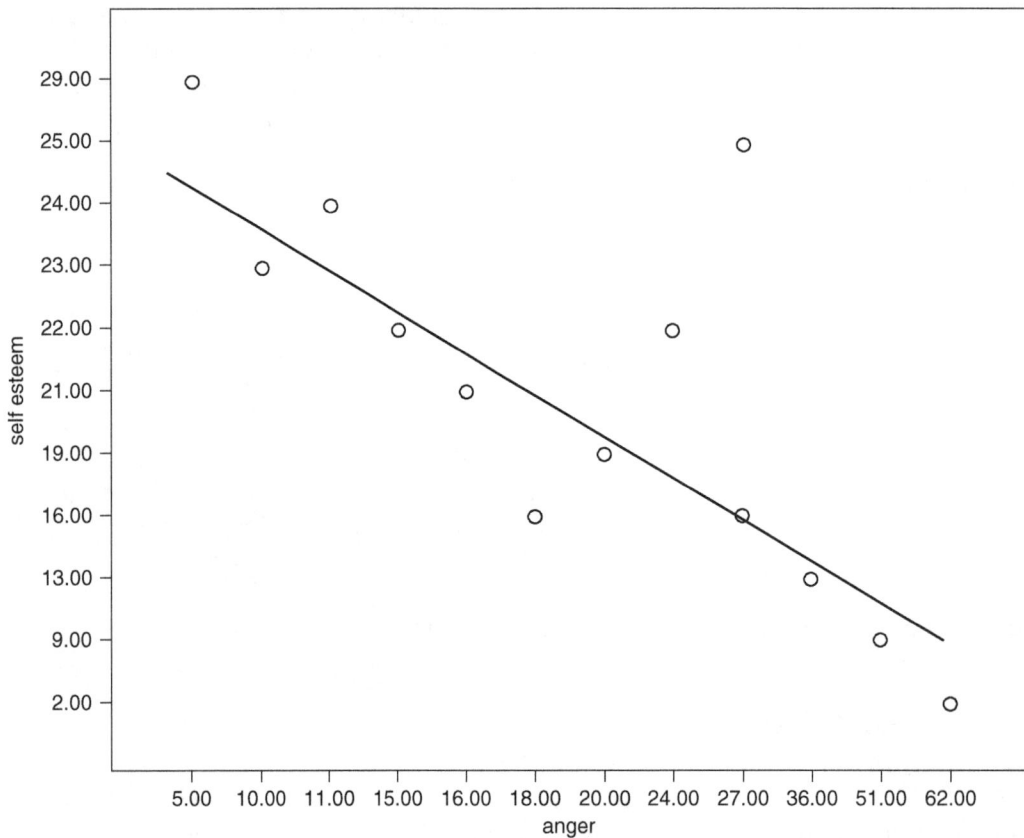

The equation becomes $\hat{Y} = 0X + \overline{Y}$. Therefore, when the correlation is 0, if someone gives you their X value, then the best estimate would be the mean of Y. The line of best fit would be a horizontal line intercepting the mean of Y.

In the simple linear regression model, we try to predict Y from X. However, most times there are multiple predictors for predicting Y. This is what is called **multiple regression**. An example of multiple regression would be to use body mass index, heart rate, pulse rate, and age to predict a health score developed by your insurance company (0 being poor health; 100 excellent health). The statistic used in multiple regression is R. R is the **multiple correlation** or strength of relationship between the dependent variable or criterion (Y variable) and the predictors (or the X variables). The value of R ranges from 0 to 1.0. There are no negative values here unlike r. In fact, as the number of predictors increase, then the value of R will go up. However, this value will increase by chance alone. Hence, you want a few independent predictors in order to optimize the value of R. In other words, you want a lean, mean, regression equation (lots of sample size and a few excellent predictors that are uncorrelated with each other). The computation and intricacies of multiple regression is a topic beyond our scope in an undergraduate course.

Correlation—The Heuristic and Computational Formulas—Class Example

A clinical psychologist was interested in determining if there was a linear relationship between age and lethal behaviors. A 22-item lethal behaviors scale (LBS) by Thorson and Powell (1987) was used (e.g., Do you own a gun? Do you like to mountain climb? Do you drive more than 65 mph?). The scores on the LBS ranged from 22 (low lethal behaviors) to 66 (high lethal behaviors). There were a total of 15 subjects in this study.

The hypothetical data set is provided below.

Subject	Age(X)	LBS(Y)	Z_x	Z_y	$Z_x Z_y$
1	21	49	−.742	1.810	−1.3440
2	34	30	.135	−.492	−.0665
3	30	28	−.135	−.735	.0992
4	18	39	−.945	.597	−.5651
5	17	36	−1.012	.234	−.2372
6	47	25	1.012	−1.098	−1.1127
7	52	28	1.350	−.735	−.9927
8	23	34	−.607	−.008	.0049
9	19	51	−.877	2.052	−1.8011
10	21	46	−.742	1.446	−1.0740
11	70	24	2.565	−1.220	−3.1299
12	47	30	1.012	−.492	−.4991
13	26	32	−.405	−.250	.1014
14	28	31	−.270	−.371	.1003
15	27	28	−.337	−.735	.2481
Total	480	511			−10.2688
SD (population)	14.81	8.25			
SD (sample)	15.33	8.53			
Mean	32.00	34.06			

1. What is the null hypothesis?

2. How do you compute a correlation (heuristically and computationally) and test it for significance?

   a. F-test

   b. z-test

3. Draw the appropriate conclusion.

4. Find the regression equation and the regression line.

5. How would you find the confidence interval for this correlation?

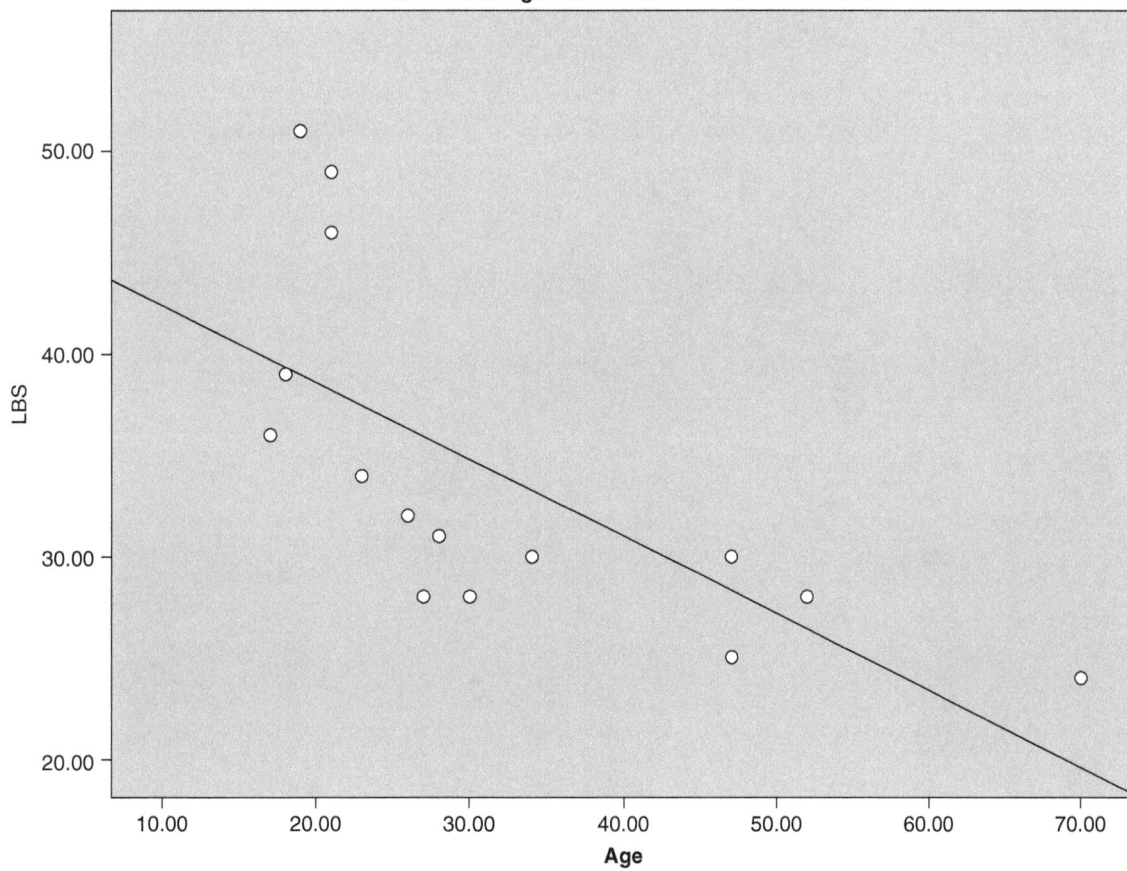

Scatter Plot of Age with Lethal Behavior Scale Scores

Homework 9

Correlation

The local Chamber of Commerce higher-ups were interested in determining if there was a correlation between the satisfaction ratings of the Electric Daisy Carnival (EDC) with that of Rock in Rio (RR). Each person attended both festivals and rated their experience on a scale of 0 (horrible) to 100 (spectacular). Moreover, they also wanted to predict the satisfaction ratings of RR from the satisfaction ratings from the EDC. Here are the hypothetical data:

EDC	RR
48	36
40	30
59	64
55	48
44	31
40	82
38	20
40	51
30	17
95	90

1. Compute the correlation, both heuristically and computationally.

2. Perform both the F- and z-tests and provide the conclusion.

3. Compute the 95% confidence interval for the correlation.

4. Provide the linear regression equation.

5. Suppose in an earlier study, the correlation of EDC and RR ratings for the 30 bikers was .35. For this group (UNLV students), test the difference between the independent correlations.

6. Provide the conclusion from the analysis.

Point-biserial correlation

A **point-biserial correlation** is when one of the variables is a true dichotomy and the other is continuous. A truly dichotomous variable is one that is automatically or naturally divided into two categories. Examples of truly dichotomous variables would be sex (male or female), race (Asian or Hispanic), or smoking (either you do or you do not). In contrast, the **biserial correlation** is when the dichotomous variable is artificial (e.g., tall–short, pass–fail, old–young) and the other variable is continuous. For instance, if you asked Billy Barty, a film actor who stood at 3 feet and 9 inches tall, then perhaps everyone above 4 feet would be considered tall and everyone below him in physical stature (e.g., younger children) would be considered short. However, if you ask DeMarcus Cousins, a basketball player who stands at 6 foot 11 inches, then perhaps everyone seven feet and above would be tall, and everyone else would be short. Thus, the artificial dichotomy depends on your point of view.

In the point-biserial correlation, suppose that we are interested in determining the relationship between the type of nurse (either LPN—licensed practical nurse, which will be coded as a 0; or RN—registered nurse, which is coded as a 1) and job satisfaction ratings. The job satisfaction ratings range from 0 (no job satisfaction) to 10 (extremely satisfied). The data are presented in Table 11.4.

Table 11.4 Hypothetical data indicating the type of nurse and their job satisfaction ratings

Nurse Type	Satisfaction
0	3
0	2
0	1
0	1
0	4
0	3
1	7
1	6
1	8
1	8
1	9
1	5

In order to obtain the point-biserial correlation (r_{pb}), use the same computational formula for correlation as we showed earlier.

$$r_{pb} = \frac{N\Sigma XY - \Sigma X \Sigma Y}{\sqrt{[N\Sigma X^2 - (\Sigma X)^2][N\Sigma Y^2 - (\Sigma Y)^2]}}$$

$$\frac{12(43) - (6)(57)}{\sqrt{[12(6) - (6)^2][12(359) - (57)^2]}} = .891$$

If we plot this point distribution, as illustrated in Figure 11.10, then as we move from LPN to RN (an increase in nursing training), the job satisfaction scores increase as well.

Figure 11.10 Point distribution of nursing type and job satisfaction scores

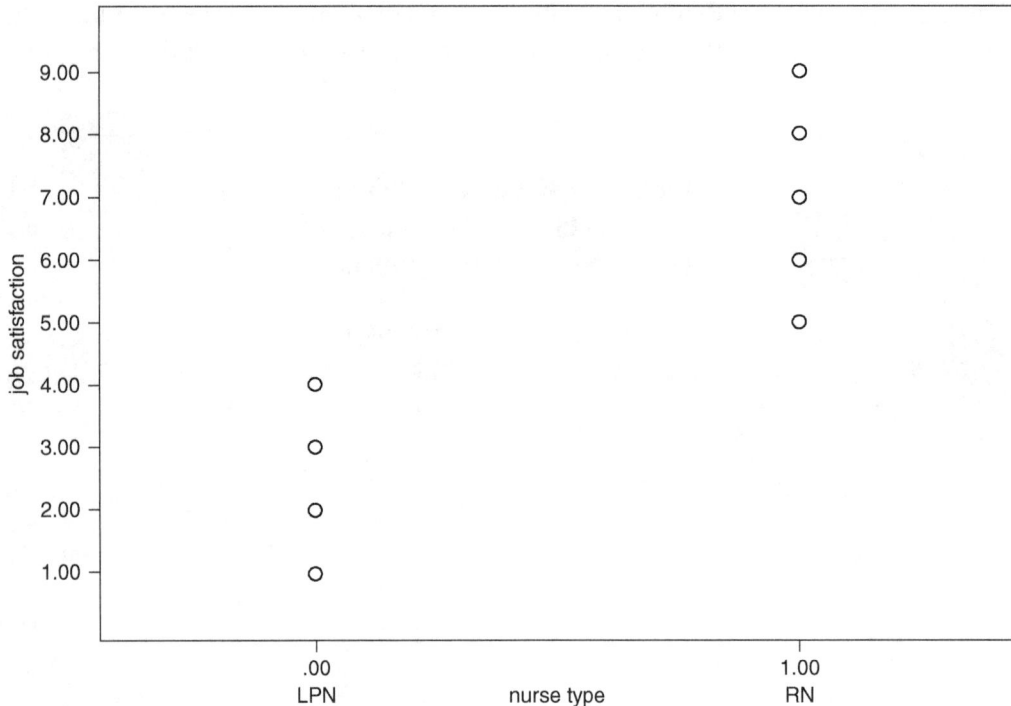

Yet, did you know that we can relate the point-biserial correlation to F?

We can think of this as a one-way ANOVA between groups with the two levels of nursing type as the independent variable and the job satisfaction ratings as the dependent variable. There would be six subjects in each group. Here is how the data would look as illustrated in Table 11.5.

**Table 11.5 Hypothetical data indicating
job satisfaction ratings for
the two different nursing types**

LPN	RN
3	7
2	6
1	8
1	8
4	9
3	5
$\Sigma X = 14$	$\Sigma X = 43$

$$\text{SS nursing type} = \frac{14^2 + 43^2}{6} - \frac{(57)^2}{12} = 70.083$$

SS within = SS total – SS nursing type

$$88.25 - 70.083 = 18.167$$

$$\text{SS total} = 359 - \frac{(57)^2}{12} = 88.25$$

Here is the summary table as illustrated in Table 11.6.

**Table 11.6 Summary table for the one-way ANOVA
of satisfaction ratings as a function of nursing types**

Source	df	SS	MS	F	p
Nursing Type	1	70.083	70.083	38.578	p < .01
Within	10	18.167	1.8167		
Total	11	88.25			

This indicates that RNs have significantly higher job satisfaction scores than do LPNs. Both the correlation and the ANOVA are statistically significant. This is not a coincidence. In fact, if you found the mean of the points for LPNs on the graph and the mean for the points for RNs and drew the line of best fit through them, in essence you would have the main effect of nursing type. Here is how the point-biserial correlation relates to F:

$$r_{pb}^2 = \frac{F}{F + df\,\text{within}}$$

Using our example

$$.891^2 = \frac{38.578}{10+38.578}$$

r_{pb}^2 is the proportion of job satisfaction variance accounted for by knowing the nursing type. The F ratio represents the amount of between (nursing type) variance in relation to the error variance. Hence, we are comparing variances.

Final Thought

We have addressed two of the most commonly used statistics: ANOVA and correlation. Yet, we have only scratched the surface of each. There are numerous other statistical procedures including multivariate (multiple dependent variables) and nonparametric (no parameters like μ or ρ involved) techniques, meta-analysis (synthesizing multiple studies allowing one to compare, contrast, and draw conclusions), and chi-square (which tests a number of different null hypotheses, such as does the observed frequency equal the expected frequency or is there a difference between population proportions?). All of these and other procedures allow for a wider variety of questions to be explored. Of course, with statistical software, the results can be obtained with drag and drops, point and clicks, or the simple entering of data. Nevertheless, one must still have a solid foundation in these techniques in order to use the software properly and interpret the results appropriately. As you begin to examine and conduct research in your discipline, allow your question to dictate the statistic as opposed to the statistic dictating your question.

References

Ahad, N. A., Yin, T. S., Othman, A. R., & Yaacob, C. R. (2011). Sensitivity of normality tests to non-normal data. *Sains Malaysiana, 40* (6) 637–641.

Brown, M. B., & Forsythe, A. B. (1974). Robust tests for equality of variances. *Journal of the American Statistical Association, 69,* 364–367.

Brysbaert, M. (1991). Algorithms for randomness in the behavioral sciences: A tutorial. *Behavior Research Methods, Instruments, & Computers, 23*(1), 45–60.

Christie, R. & Geis, F. (1970). *Studies in Machiavellianism*. New York: Academic Press.

Conover, W. J., Johnson, M. E., & Johnson, M. M. (1981). A comparative study of tests for homogeneity of variances, with applications to the outer continental shelf bidding data. *Technometrics, 23*(4), 351–361.

Duncan, D. B. (1955). Multiple range and multiple F tests. *Biometrics, 11*(1), 1–42.

Edgell, S. E., & Noon, S. M. (1984). Effect of violation of normality on the t test of the correlation coefficient. *Psychological Bulletin, 95*(3), 576–583.

Fisher, R. A. (1921). On the probable error of a coefficient of a correlation deduced from a small sample. *Metron, 1,* 1–32.

Gardner, W. M., & Melvin, K. B. (1988). A scale for measuring attitude toward cheating. *Bulletin of the Psychonomic Society, 26*(5), 429–432.

Havlicek, L. L., & Peterson, N. L. (1977). Effect of the violation of assumptions upon significance levels of the Pearson r. *Psychological Bulletin, 84*(2), 373–377.

Hayter, A. J. (1986). The maximum familywise error rate of Fisher's least significant difference test. *Journal of the American Statistical Association, 81,* 1000–1004.

Howell, D. C. (2010). *Statistical methods for psychology* (8th ed). New York: Cengage.

Huck, S. W., & Malgady, R. G. (1978). Two-way analysis of variance using means and standard deviations. *Educational and Psychological Measurement, 38*(2), 235–237.

Kenward, M.G, & Roger J.H. (1997). Small sample inference for fixed effects from restricted maximum likelihood. *Biometrics, 53*(3), 983–997 doi: 10.2307/2533558.

Keuls, M. (1952). The use of the "studentized range" in connection with an analysis of variance. *Euphytica, 1*(2), 112–122.

Madden, R. A., & Williams, J. (1978). The correlation between temperature and precipitation in the United States and Europe. *Monthly Weather Review, 106*(1), 142–147.

Martin, S. A., & Toothaker, L. E. (1989). PERITZ: A FORTRAN program for performing multiple comparisons of means using the Peritz Q method. *Behavior Research Methods, 21*(4), 465–472.

Newman, D. (1939). The distribution of range in samples from a normal population, expressed in terms of an independent estimate of standard deviation. *Biometrika, 31,* 20–30.

Petrinovich, L. F., & Hardyck, C. D. (1969). Error rates for multiple comparison methods: Some evidence concerning the frequency of erroneous conclusions. *Psychological Bulletin, 71*(1), 43–54.

Rosenberg, M. (1965). *Society and the adolescent self-image.* Princeton, NJ: Princeton University Press.

Satterthwaite, F. E. (1941). Synthesis of variance. *Psychometrika, 16*(5), 309–316.

Satterthwaite, F. E. (1946). An approximate distribution of estimates of variance components. *Biometrics Bulletin, 2,* 110–114.

Schaalje, G.B., McBride, J.B, & Fellingham, G.W. (2002). Adequacy of approximations to distributions of test statistics in complex mixed linear models. *Journal of Agricultural, Biological, and Environmental Statistics, 7*(4), 512–524. doi: 10.1198/108571102726.

Silver, N. C., & Dunlap, W. P. (1987). Averaging correlation coefficients: Should Fisher's z' transformation be used? *Journal of Applied Psychology, 72*(1), 146–148.

Snell, W. E., Gum, S., Shuck, R. L., Mosley, J. A., & Kite, T. L. (1995). The Clinical Anger Scale: preliminary reliability and validity. *Journal of Clinical Psychology, 51*(2), 215–226.

Thorson, J. A. & Powell, F.C. (1987). Factor stucture of the Lethal Behaviors Scale. *Psychological Reports, 61* (3), 807–810.

Tukey, J. W. (1953). *The problem of multiple comparisons.* Unpublished manuscript. Princeton University.

Tulloh, B. R., & Collopy, B. T. (1994). Positive correlation between blood alcohol level and ISS in road trauma. *Injury, 25*(8), 539–543.

F table for $\alpha = 0.05$

df_2 \ df_1	1	2	3	4	5	6	7	8	9	10	12	15	20	24	30	40	60	120	∞
1	161.4476	199.5000	215.7073	224.5832	230.1619	233.9860	236.7684	238.8827	240.5433	241.8817	243.9060	245.9499	248.0131	249.0518	250.0951	251.1432	252.1957	253.2529	254.3144
2	18.5128	19.0000	19.1643	19.2468	19.2964	19.3295	19.3532	19.3710	19.3848	19.3959	19.4125	19.4291	19.4458	19.4541	19.4624	19.4707	19.4791	19.4874	19.4957
3	10.1280	9.5521	9.2766	9.1172	9.0135	8.9406	8.8867	8.8452	8.8123	8.7855	8.7446	8.7029	8.6602	8.6385	8.6166	8.5944	8.5720	8.5494	8.5264
4	7.7086	6.9443	6.5914	6.3882	6.2561	6.1631	6.0942	6.0410	5.9988	5.9644	5.9117	5.8578	5.8025	5.7744	5.7459	5.7170	5.6877	5.6581	5.6281
5	6.6079	5.7861	5.4095	5.1922	5.0503	4.9503	4.8759	4.8183	4.7725	4.7351	4.6777	4.6188	4.5581	4.5272	4.4957	4.4638	4.4314	4.3985	4.3650
6	5.9874	5.1433	4.7571	4.5337	4.3874	4.2839	4.2067	4.1468	4.0990	4.0600	3.9999	3.9381	3.8742	3.8415	3.8082	3.7743	3.7398	3.7047	3.6689
7	5.5914	4.7374	4.3468	4.1203	3.9715	3.8660	3.7870	3.7257	3.6767	3.6365	3.5747	3.5107	3.4445	3.4105	3.3758	3.3404	3.3043	3.2674	3.2298
8	5.3177	4.4590	4.0662	3.8379	3.6875	3.5806	3.5005	3.4381	3.3881	3.3472	3.2839	3.2184	3.1503	3.1152	3.0794	3.0428	3.0053	2.9669	2.9276
9	5.1174	4.2565	3.8625	3.6331	3.4817	3.3738	3.2927	3.2296	3.1789	3.1373	3.0729	3.0061	2.9365	2.9005	2.8637	2.8259	2.7872	2.7475	2.7067
10	4.9646	4.1028	3.7068	3.4780	3.3258	3.2172	3.1355	3.0717	3.0204	2.9782	2.9130	2.8450	2.7740	2.7372	2.6996	2.6609	2.6211	2.5801	2.5379
11	4.8443	3.9823	3.5874	3.3567	3.2039	3.0946	3.0123	2.9480	2.8962	2.8536	2.7876	2.7186	2.6464	2.6090	2.5705	2.5309	2.4901	2.4480	2.4045
12	4.7472	3.8853	3.4903	3.2592	3.1059	2.9961	2.9134	2.8486	2.7964	2.7534	2.6866	2.6169	2.5436	2.5055	2.4663	2.4259	2.3842	2.3410	2.2962
13	4.6672	3.8056	3.4105	3.1791	3.0254	2.9153	2.8321	2.7669	2.7144	2.6710	2.6037	2.5331	2.4589	2.4202	2.3803	2.3392	2.2966	2.2524	2.2064
14	4.6001	3.7389	3.3439	3.1122	2.9582	2.8477	2.7642	2.6987	2.6458	2.6022	2.5342	2.4630	2.3879	2.3487	2.3082	2.2664	2.2229	2.1778	2.1307
15	4.5431	3.6823	3.2874	3.0556	2.9013	2.7905	2.7066	2.6408	2.5876	2.5437	2.4753	2.4034	2.3275	2.2878	2.2468	2.2043	2.1601	2.1141	2.0658
16	4.4940	3.6337	3.2389	3.0069	2.8524	2.7413	2.6572	2.5911	2.5377	2.4935	2.4247	2.3522	2.2756	2.2354	2.1938	2.1507	2.1058	2.0589	2.0096
17	4.4513	3.5915	3.1968	2.9647	2.8100	2.6987	2.6143	2.5480	2.4943	2.4499	2.3807	2.3077	2.2304	2.1898	2.1477	2.1040	2.0584	2.0170	1.9604
18	4.4139	3.5546	3.1599	2.9277	2.7729	2.6613	2.5767	2.5102	2.4563	2.4117	2.3421	2.2686	2.1906	2.1497	2.1071	2.0629	2.0166	1.9681	1.9168
19	4.3807	3.5219	3.1274	2.8951	2.7401	2.6283	2.5435	2.4768	2.4227	2.3779	2.3080	2.2341	2.1555	2.1141	2.0712	2.0264	1.9795	1.9302	1.8780
20	4.3512	3.4928	3.0984	2.8661	2.7109	2.5990	2.5140	2.4471	2.3928	2.3479	2.2776	2.2033	2.1242	2.0825	2.0391	1.9938	1.9464	1.8963	1.8432
21	4.3248	3.4668	3.0725	2.8401	2.6848	2.5727	2.4876	2.4205	2.3660	2.3210	2.2504	2.1757	2.0960	2.0540	2.0102	1.9645	1.9165	1.8657	1.8117
22	4.3009	3.4434	3.0491	2.8167	2.6613	2.5491	2.4638	2.3965	2.3419	2.2967	2.2258	2.1508	2.0707	2.0283	1.9842	1.9380	1.8894	1.8380	1.7831
23	4.2793	3.4221	3.0280	2.7955	2.6400	2.5277	2.4422	2.3748	2.3201	2.2747	2.2036	2.1282	2.0476	2.0050	1.9605	1.9139	1.8648	1.8128	1.7570
24	4.2597	3.4028	3.0088	2.7763	2.6207	2.5082	2.4226	2.3551	2.3002	2.2547	2.1834	2.1077	2.0267	1.9838	1.9390	1.8920	1.8424	1.7896	1.7330
25	4.2417	3.3852	2.9912	2.7587	2.6030	2.4904	2.4047	2.3371	2.2821	2.2365	2.1649	2.0889	2.0075	1.9643	1.9192	1.8718	1.8217	1.7684	1.7110
26	4.2252	3.3690	2.9752	2.7426	2.5868	2.4741	2.3883	2.3205	2.2655	2.2197	2.1479	2.0716	1.9898	1.9464	1.9010	1.8533	1.8027	1.7483	1.6906
27	4.2100	3.3541	2.9604	2.7278	2.5719	2.4591	2.3732	2.3053	2.2501	2.2043	2.1323	2.0558	1.9736	1.9299	1.8842	1.8361	1.7851	1.7306	1.6717
28	4.1960	3.3404	2.9467	2.7141	2.5581	2.4453	2.3593	2.2913	2.2360	2.1900	2.1179	2.0411	1.9586	1.9147	1.8687	1.8203	1.7689	1.7138	1.6541
29	4.1830	3.3277	2.9340	2.7014	2.5454	2.4324	2.3463	2.2783	2.2229	2.1768	2.1045	2.0275	1.9446	1.9005	1.8453	1.8055	1.7537	1.6981	1.6376
30	4.1709	3.3158	2.9223	2.6896	2.5336	2.4205	2.3343	2.2662	2.2107	2.1646	2.0921	2.0148	1.9317	1.8874	1.8409	1.7918	1.7396	1.6835	1.6223
40	4.0847	3.2317	2.8387	2.6060	2.4495	2.3359	2.2490	2.1802	2.1240	2.0772	2.0035	1.9245	1.8389	1.7929	1.7444	1.6928	1.6373	1.5766	1.5089
60	4.0012	3.1504	2.7581	2.5252	2.3683	2.2541	2.1665	2.0970	2.0401	1.9926	1.9174	1.8364	1.7480	1.7001	1.6491	1.5943	1.5343	1.4673	1.3893
120	3.9201	3.0718	2.6802	2.4472	2.2899	2.1750	2.0868	2.0164	1.9588	1.9105	1.8337	1.7505	1.6587	1.6084	1.5543	1.4952	1.4290	1.3519	1.2539
∞	3.8415	2.9957	2.6049	2.3719	2.2141	2.0986	2.0096	1.9384	1.8799	1.8307	1.7522	1.6664	1.5705	1.5173	1.4591	1.3940	1.3180	1.2214	1.0000

F table for α = 0.01

df_2 \ df_1	1	2	3	4	5	6	7	8	9	10	12	15	20	24	30	40	60	120	∞
1	4052.181	4999.500	5403.352	5624.583	5763.650	5858.986	5928.356	5981.070	6022.473	6055.847	6106.321	6157.285	6208.730	6234.631	6260.649	6286.782	6313.030	6339.391	6365.864
2	98.503	99.000	99.166	99.249	99.299	99.333	99.356	99.374	99.388	99.399	99.416	99.433	99.449	99.458	99.466	99.474	99.482	99.491	99.499
3	34.116	30.817	29.457	28.710	28.237	27.911	27.672	27.489	27.345	27.229	27.052	26.872	26.690	26.598	26.505	26.411	26.316	26.221	26.125
4	21.198	18.000	16.694	15.977	15.522	15.207	14.976	14.799	14.659	14.546	14.374	14.198	14.020	13.929	13.838	13.745	13.652	13.558	13.463
5	16.258	13.274	12.060	11.392	10.967	10.672	10.456	10.289	10.158	10.051	9.888	9.722	9.553	9.466	9.379	9.291	9.202	9.112	9.020
6	13.745	10.925	9.780	9.148	8.746	8.466	8.260	8.102	7.976	7.874	7.718	7.559	7.396	7.313	7.229	7.143	7.057	6.969	6.880
7	12.246	9.547	8.451	7.847	7.460	7.191	6.993	6.840	6.719	6.620	6.469	6.314	6.155	6.074	5.992	5.908	5.824	5.737	5.650
8	11.259	8.649	7.591	7.006	6.632	6.371	6.178	6.029	5.911	5.814	5.667	5.515	5.359	5.279	5.198	5.116	5.032	4.946	4.859
9	10.561	8.022	6.992	6.422	6.057	5.802	5.613	5.467	5.351	5.257	5.111	4.962	4.808	4.729	4.649	4.567	4.483	4.398	4.311
10	10.044	7.559	6.552	5.994	5.636	5.386	5.200	5.057	4.942	4.849	4.706	4.558	4.405	4.327	4.247	4.165	4.082	3.996	3.909
11	9.646	7.206	6.217	5.668	5.316	5.069	4.886	4.744	4.632	4.539	4.397	4.251	4.099	4.021	3.941	3.860	3.776	3.690	3.602
12	9.330	6.927	5.953	5.412	5.064	4.821	4.640	4.499	4.388	4.296	4.155	4.010	3.858	3.780	3.701	3.619	3.535	3.449	3.361
13	9.074	6.701	5.739	5.205	4.862	4.620	4.441	4.302	4.191	4.100	3.960	3.815	3.665	3.587	3.507	3.425	3.341	3.255	3.165
14	8.862	6.515	5.564	5.035	4.695	4.456	4.278	4.140	4.030	3.939	3.800	3.656	3.505	3.427	3.348	3.266	3.181	3.094	3.004
15	8.683	6.359	5.417	4.893	4.556	4.318	4.142	4.004	3.895	3.805	3.666	3.522	3.372	3.294	3.214	3.132	3.047	2.959	2.868
16	8.531	6.226	5.292	4.773	4.437	4.202	4.026	3.890	3.780	3.691	3.553	3.409	3.259	3.181	3.101	3.018	2.933	2.845	2.753
17	8.400	6.112	5.185	4.669	4.336	4.102	3.927	3.791	3.682	3.593	3.455	3.312	3.162	3.084	3.003	2.920	2.835	2.746	2.653
18	8.285	6.013	5.092	4.579	4.243	4.015	3.841	3.705	3.597	3.508	3.371	3.227	3.077	2.999	2.919	2.835	2.749	2.660	2.566
19	8.185	5.926	5.010	4.500	4.171	3.939	3.765	3.631	3.523	3.434	3.297	3.153	3.003	2.925	2.844	2.761	2.674	2.584	2.489
20	8.096	5.849	4.938	4.431	4.103	3.871	3.699	3.564	3.457	3.368	3.231	3.088	2.938	2.859	2.778	2.695	2.608	2.517	2.421
21	8.017	5.780	4.874	4.369	4.042	3.812	3.640	3.506	3.398	3.310	3.173	3.030	2.880	2.801	2.720	2.636	2.548	2.457	2.360
22	7.945	5.719	4.817	4.313	3.988	3.758	3.587	3.453	3.346	3.258	3.121	2.978	2.827	2.749	2.667	2.583	2.495	2.403	2.305
23	7.881	5.664	4.765	4.264	3.939	3.710	3.539	3.406	3.299	3.211	3.074	2.931	2.781	2.702	2.620	2.535	2.447	2.354	2.256
24	7.823	5.614	4.718	4.218	3.895	3.667	3.496	3.363	3.256	3.168	3.032	2.889	2.738	2.659	2.577	2.492	2.403	2.310	2.211
25	7.770	5.568	4.675	4.177	3.855	3.627	3.457	3.324	3.217	3.129	2.993	2.850	2.699	2.620	2.538	2.453	2.364	2.270	2.169
26	7.721	5.526	4.637	4.140	3.818	3.591	3.421	3.288	3.182	3.094	2.958	2.815	2.664	2.585	2.503	2.417	2.327	2.233	2.131
27	7.677	5.488	4.601	4.106	3.785	3.558	3.388	3.256	3.149	3.062	2.926	2.783	2.632	2.552	2.470	2.384	2.294	2.198	2.097
28	7.636	5.453	4.568	4.074	3.754	3.528	3.358	3.225	3.120	3.032	2.896	2.753	2.602	2.522	2.440	2.354	2.263	2.167	2.064
29	7.598	5.420	4.538	4.045	3.725	3.499	3.330	3.198	3.092	3.005	2.868	2.726	2.574	2.495	2.412	2.325	2.234	2.138	2.034
30	7.562	5.390	4.510	4.018	3.699	3.473	3.304	3.173	3.067	2.979	2.843	2.700	2.549	2.469	2.386	2.299	2.208	2.111	2.006
40	7.314	5.179	4.313	3.828	3.514	3.291	3.124	2.993	2.888	2.801	2.665	2.522	2.369	2.288	2.203	2.114	2.019	1.917	1.805
80	7.077	4.977	4.126	3.649	3.339	3.119	2.953	2.823	2.718	2.632	2.496	2.352	2.198	2.115	2.028	1.936	1.836	1.726	1.601
120	6.851	4.787	3.949	3.480	3.174	2.956	2.792	2.663	2.559	2.472	2.336	2.192	2.035	1.950	1.860	1.763	1.656	1.533	1.381
∞	6.635	4.605	3.782	3.319	3.017	2.802	2.639	2.511	2.407	2.321	2.185	2.039	1.878	1.791	1.696	1.592	1.473	1.325	1.000

Critical values for Q

df for Error Term	k = Number of Treatments								
	2	3	4	5	6	7	8	9	10
5	3.64	4.60	5.22	5.67	6.03	6.33	6.53	6.30	6.99
	5.70	6.98	7.80	8.42	8.91	9.32	9.67	9.67	10.24
6	3.46	4.34	4.90	5.30	5.63	5.90	6.12	6.12	6.49
	5.24	6.33	7.03	7.56	7.97	8.32	8.61	8.87	9.10
7	3.34	4.16	4.68	5.06	5.36	5.61	5.82	6.00	6.16
	4.95	5.92	6.54	7.01	7.37	7.68	7.94	8.17	8.37
8	3.26	4.04	4.53	4.89	5.17	5.40	5.60	5.77	5.92
	4.75	5.64	6.20	6.62	6.96	7.24	7.47	7.68	7.86
9	3.20	3.95	4.41	4.76	5.02	5.24	5.43	5.59	5.74
	4.60	5.43	5.96	6.35	6.66	6.91	7.13	7.33	7.49
10	3.15	3.88	4.33	4.65	4.91	5.12	5.30	5.46	5.60
	4.48	5.27	5.77	6.14	6.43	6.67	6.87	7.05	7.21
11	3.11	3.82	4.26	4.57	4.32	5.03	5.20	5.35	5.49
	4.39	5.15	5.62	5.97	6.25	6.48	6.67	6.84	6.99
12	3.08	3.77	4.20	4.51	4.75	4.95	5.12	5.27	5.39
	4.32	5.05	5.50	5.84	6.10	6.32	6.51	6.67	6.81
13	3.06	3.73	4.15	4.45	4.69	4.33	5.05	5.19	5.32
	4.26	4.96	5.40	5.73	5.98	6.19	6.37	6.53	6.67
14	3.03	3.70	4.11	4.41	4.64	4.83	4.99	5.13	5.25
	4.21	4.89	5.32	5.63	5.88	6.08	6.26	6.41	6.54
15	3.01	3.67	4.08	4.37	4.59	4.78	4.94	5.08	5.20
	4.17	4.84	5.25	5.56	5.80	5.99	6.16	6.31	6.44
16	3.00	3.65	4.05	4.33	4.56	4.74	4.90	5.03	5.15
	4.13	4.79	5.19	5.49	5.72	5.92	6.08	6.22	6.35
17	2.98	3.63	4.02	4.30	4.52	4.70	4.86	4.99	5.11
	4.10	4.74	5.14	5.43	5.66	5.85	6.01	6.15	6.27
18	2.97	3.61	4.00	4.28	4.49	4.67	4.82	4.96	5.07
	4.07	4.70	5.09	5.38	5.60	5.79	5.94	6.08	6.20
19	2.96	3.59	3.98	4.25	4.47	4.65	4.79	4.92	5.04
	4.05	4.67	5.05	5.33	5.55	5.73	5.89	6.02	6.14
20	2.95	3.58	3.96	4.23	4.45	4.62	4.77	4.90	5.01
	4.02	4.64	5.02	5.29	5.51	5.69	5.84	5.97	6.09
24	2.92	3.53	3.90	4.17	4.37	4.54	4.68	4.81	4.92
	3.96	4.55	4.91	5.17	5.37	5.54	5.69	5.81	5.92
30	2.89	3.49	3.85	4.10	4.30	4.46	4.60	4.72	4.82
	3.89	4.45	4.80	5.05	5.24	5.40	5.54	5.65	5.76
40	2.86	3.44	3.79	4.04	4.23	4.39	4.52	4.63	4.73
	3.82	4.37	4.70	4.93	5.11	5.26	5.39	5.50	5.60
60	2.83	3.40	3.74	3.98	4.16	4.31	4.44	4.55	4.65
	3.76	4.28	4.59	4.82	4.99	5.13	5.25	5.36	5.45
120	2.30	3.36	3.68	3.92	4.10	4.24	4.36	4.47	4.56
	3.70	4.20	4.50	4.71	4.87	5.01	5.12	5.21	5.30
infinity	2.77	3.31	3.63	3.86	4.03	4.17	4.29	4.39	4.47
	3.64	4.12	4.40	4.60	4.76	4.88	4.99	5.08	5.16

Pearson r to Fisher z′

r	z′	r	z′	r	z′
0.00	0.0000	0.41	0.4356	0.81	1.1270
0.01	0.0100	0.42	0.4477	0.82	1.1568
0.02	0.0200	0.43	0.4599	0.83	1.1881
0.03	0.0300	0.44	0.4722	0.84	1.2212
0.04	0.0400	0.45	0.4847	0.85	1.2562
0.05	0.0500	0.46	0.4973	0.86	1.2933
0.06	0.0601	0.47	0.5101	0.87	1.3331
0.07	0.0701	0.48	0.5230	0.88	1.3758
0.08	0.0802	0.49	0.5361	0.89	1.4219
0.09	0.0902	0.50	0.5493	0.90	1.4722
0.10	0.1003	0.51	0.5627	0.91	1.5275
0.11	0.1104	0.52	0.5763	0.92	1.5890
0.12	0.1206	0.53	0.5901	0.93	1.6584
0.13	0.1307	0.54	0.6042	0.94	1.7380
0.14	0.1409	0.55	0.6184	0.95	1.8318
0.15	0.1511	0.56	0.6328	0.96	1.9459
0.16	0.1614	0.57	0.6475	0.97	2.0923
0.17	0.1717	0.58	0.6625	0.98	2.2976
0.18	0.1820	0.59	0.6777	0.99	2.6467
0.19	0.1923	0.60	0.6931		
0.20	0.2027	0.61	0.7089		
0.21	0.2132	0.62	0.7250		
0.22	0.2237	0.63	0.7414		
0.23	0.2342	0.64	0.7582		
0.24	0.2448	0.65	0.7753		
0.25	0.2554	0.66	0.7928		
0.26	0.2661	0.67	0.8107		
0.27	0.2769	0.68	0.8291		
0.28	0.2877	0.69	0.8480		
0.29	0.2986	0.70	0.8673		
0.30	0.3095	0.71	0.8872		
0.31	0.3205	0.72	0.9076		
0.32	0.3316	0.73	0.9287		
0.33	0.3428	0.74	0.9505		
0.34	0.3541	0.75	0.9730		
0.35	0.3654	0.76	0.9962		
0.36	0.3769	0.77	1.0203		
0.37	0.3884	0.78	1.0454		
0.38	0.4001	0.79	1.0714		
0.39	0.4118	0.80	1.0986		
0.40	0.4236				

The Normal Distribution (z)

z	Mean to z	z to tail	z	Mean to z	z to tail
.00	.0000	.5000	.40	.1554	.3446
.01	.0040	.4960	.41	.1591	.3409
.02	.0080	.4920	.42	.1628	.3372
.03	.0120	.4880	.43	.1664	.3336
.04	.0160	.4840	.44	.1700	.3300
.05	.0199	.4801	.45	.1736	.3264
.06	.0239	.4761	.46	.1772	.3228
.07	.0279	.4721	.47	.1808	.3192
.08	.0319	.4681	.48	.1844	.3156
.09	.0359	.4641	.49	.1879	.3121
.10	.0398	.4602	.50	.1915	.3085
.11	.0438	.4562	.51	.1950	.3050
.12	.0478	.4522	.52	.1985	.3015
.13	.0517	.4483	.53	.2019	.2981
.14	.0557	.4443	.54	.2054	.2946
.15	.0596	.4404	.55	.2088	.2912
.16	.0636	.4364	.56	.2123	.2877
.17	.0675	.4325	.57	.2157	.2843
.18	.0714	.4286	.58	.2190	.2810
.19	.0753	.4247	.59	.2224	.2776
.20	.0793	.4207	.60	.2257	.2743
.21	.0832	.4168	.61	.2291	.2709
.22	.0871	.4129	.62	.2324	.2676
.23	.0910	.4090	.63	.2357	.2643
.24	.0948	.4052	.64	.2389	.2611
.25	.0987	.4013	.65	.2422	.2578
.26	.1026	.3974	.66	.2454	.2546
.27	.1064	.3936	.67	.2486	.2514
.28	.1103	.3897	.68	.2517	.2483
.29	.1141	.3859	.69	.2549	.2451
.30	.1179	.3821	.70	.2580	.2420
.31	.1217	.3783	.71	.2611	.2389
.32	.1255	.3745	.72	.2642	.2358
.33	.1293	.3707	.73	.2673	.2327
.34	.1331	.3669	.74	.2704	.2296
.35	.1368	.3632	.75	.2734	.2266
.36	.1406	.3594	.76	.2764	.2236
.37	.1443	.3557	.77	.2794	.2206
.38	.1480	.3520	.78	.2823	.2177
.39	.1517	.3483	.79	.2852	.2148

(Continues)

The Normal Distribution (z) (*Continued*)

z	Mean to z	z to tail	z	Mean to z	z to tail
.80	.2881	.2119	1.20	.3849	.1151
.81	.2910	.2090	1.21	.3869	.1131
.82	.2939	.2061	1.22	.3888	.1112
.83	.2967	.2033	1.23	.3907	.1093
.84	.2995	.2005	1.24	.3925	.1075
.85	.3023	.1977	1.25	.3944	.1056
.86	.3051	.1949	1.26	.3962	.1038
.87	.3078	.1922	1.27	.3980	.1020
.88	.3106	.1894	1.28	.3997	.1003
.89	.3133	.1867	1.29	.4015	.0985
.90	.3159	.1841	1.30	.4032	.0968
.91	.3186	.1814	1.31	.4049	.0951
.92	.3212	.1788	1.32	.4066	.0934
.93	.3238	.1762	1.33	.4082	.0918
.94	.3264	.1736	1.34	.4099	.0901
.95	.3289	.1711	1.35	.4115	.0885
.96	.3315	.1685	1.36	.4131	.0869
.97	.3340	.1660	1.37	.4147	.0853
.98	.3365	.1635	1.38	.4162	.0838
.99	.3389	.1611	1.39	.4177	.0823
1.00	.3413	.1587	1.40	.4192	.0808
1.01	.3438	.1562	1.41	.4207	.0793
1.02	.3461	.1539	1.42	.4222	.0778
1.03	.3485	.1515	1.43	.4236	.0764
1.04	.3508	.1492	1.44	.4251	.0749
1.05	.3531	.1469	1.45	.4265	.0735
1.06	.3554	.1446	1.46	.4279	.0721
1.07	.3577	.1423	1.47	.4292	.0708
1.08	.3599	.1401	1.48	.4306	.0694
1.09	.3621	.1379	1.49	.4319	.0681
1.10	.3643	.1357	1.50	.4332	.0668
1.11	.3665	.1335	1.51	.4345	.0655
1.12	.3686	.1314	1.52	.4357	.0643
1.13	.3708	.1292	1.53	.4370	.0630
1.14	.3729	.1271	1.54	.4382	.0618
1.15	.3749	.1251	1.55	.4394	.0606
1.16	.3770	.1230	1.56	.4406	.0594
1.17	.3790	.1210	1.57	.4418	.0582
1.18	.3810	.1190	1.58	.4429	.0571
1.19	.3830	.1170	1.59	.4441	.0559

(*Continues*)

The Normal Distribution (z) (*Continued*)

z	Mean to z	z to tail	z	Mean to z	z to tail
1.60	.4452	.0548	1.98	.4761	.0239
1.61	.4463	.0537	1.99	.4767	.0233
1.62	.4474	.0526	2.00	.4772	.0228
1.63	.4484	.0516	2.01	.4778	.0222
1.64	.4495	.0505	2.02	.4783	.0217
1.65	.4505	.0495	2.03	.4788	.0212
1.66	.4515	.0485	2.04	.4793	.0207
1.67	.4525	.0475	2.05	.4798	.0202
1.68	.4535	.0465	2.06	.4803	.0197
1.69	.4545	.0455	2.07	.4808	.0192
1.70	.4554	.0446	2.08	.4812	.0188
1.71	.4564	.0436	2.09	.4817	.0183
1.72	.4573	.0427	2.10	.4821	.0179
1.73	.4582	.0418	2.11	.4826	.0174
1.74	.4591	.0409	2.12	.4830	.0170
1.75	.4599	.0401	2.13	.4834	.0166
1.76	.4608	.0392	2.14	.4838	.0162
1.77	.4616	.0384	2.15	.4842	.0158
1.78	.4625	.0375	2.16	.4846	.0154
1.79	.4633	.0367	2.17	.4850	.0150
1.80	.4641	.0359	2.18	.4854	.0146
1.81	.4649	.0351	2.19	.4857	.0143
1.82	.4656	.0344	2.20	.4861	.0139
1.83	.4664	.0336	2.21	.4864	.0136
1.84	.4671	.0329	2.22	.4868	.0132
1.85	.4678	.0322	2.23	.4871	.0129
1.86	.4686	.0314	2.24	.4875	.0125
1.87	.4693	.0307	2.25	.4878	.0122
1.88	.4699	.0301	2.26	.4881	.0119
1.89	.4706	.0294	2.27	.4884	.0116
1.90	.4713	.0287	2.28	.4887	.0113
1.91	.4719	.0281	2.29	.4890	.0110
1.92	.4726	.0274	2.30	.4893	.0107
1.93	.4732	.0268	2.31	.4896	.0104
1.94	.4738	.0262	2.32	.4898	.0102
1.95	.4744	.0256	2.33	.4901	.0099
1.96	.4750	.0250	2.34	.4904	.0096
1.97	.4756	.0244	2.35	.4906	.0094

(*Continues*)

The Normal Distribution (z) (*Continued*)

z	Mean to z	z to tail	z	Mean to z	z to tail
2.36	.4909	.0091	2.72	.4967	.0033
2.37	.4911	.0089	2.73	.4968	.0032
2.38	.4913	.0087	2.74	.4969	.0031
2.39	.4916	.0084	2.75	.4970	.0030
2.40	.4918	.0082	2.76	.4971	.0029
2.41	.4920	.0080	2.77	.4972	.0028
2.42	.4922	.0078	2.78	.4973	.0027
2.43	.4925	.0075	2.79	.4974	.0026
2.44	.4927	.0073	2.80	.4974	.0026
2.45	.4929	.0071	2.81	.4975	.0025
2.46	.4931	.0069	2.82	.4976	.0024
2.47	.4932	.0068	2.83	.4977	.0023
2.48	.4934	.0066	2.84	.4977	.0023
2.49	.4936	.0064	2.85	.4978	.0022
2.50	.4938	.0062	2.86	.4979	.0021
2.51	.4940	.0060	2.87	.4979	.0021
2.52	.4941	.0059	2.88	.4980	.0020
2.53	.4943	.0057	2.89	.4981	.0019
2.54	.4945	.0055	2.90	.4981	.0019
2.55	.4946	.0054	2.91	.4982	.0018
2.56	.4948	.0052	2.92	.4982	.0018
2.57	.4949	.0051	2.93	.4983	.0017
2.58	.4951	.0049	2.94	.4984	.0016
2.59	.4952	.0048	2.95	.4984	.0016
2.60	.4953	.0047	2.96	.4985	.0015
2.61	.4955	.0045	2.97	.4985	.0015
2.62	.4956	.0044	2.98	.4986	.0014
2.63	.4957	.0043	2.99	.4986	.0014
2.64	.4959	.0041	3.00	.4987	.0013
2.65	.4960	.0040	⋮	⋮	⋮
2.66	.4961	.0039	3.25	.4994	.0006
2.67	.4962	.0038	⋮	⋮	⋮
2.68	.4963	.0037	3.50	.4998	.0002
2.69	.4964	.0036	⋮	⋮	⋮
2.70	.4965	.0035	3.75	.4999	.0001
2.71	.4966	.0034	⋮	⋮	⋮
			4.00	.4999	.0000

Critical Values for the t-Distribution

Degrees of Freedom	Levels of Significance for a One-Tailed Test					
	0.25	0.05	0.025	0.01	0.005	0.0005
	Levels of Significance for a Two-Tailed Test					
	0.50	0.10	0.05	0.02	0.01	0.001
1	1.000	6.314	12.706	31.821	63.657	636.619
2	.816	2.920	4.303	6.965	9.925	31.598
3	.765	2.353	3.182	4.541	5.841	12.924
4	.741	2.132	2.775	3.747	4.604	8.610
5	.727	2.015	2.571	3.365	4.032	6.869
6	.718	1.943	2.447	3.143	3.707	5.959
7	.711	1.895	2.365	2.998	3.499	5.408
8	.706	1.860	2.306	2.896	3.355	5.041
9	.703	1.833	2.262	2.821	3.250	4.781
10	.700	1.812	2.228	2.764	3.169	4.587
11	.697	1.796	2.201	2.718	3.106	4.437
12	.695	1.782	2.179	2.681	3.055	4.318
13	.694	1.771	2.160	2.650	3.012	4.221
14	.692	1.761	2.145	2.624	2.977	4.140
15	.691	1.753	2.131	2.602	2.947	4.073
16	.690	1.746	2.120	2.583	2.921	4.015
17	.689	1.740	2.110	2.567	2.898	3.965
18	.688	1.734	2.101	2.552	2.878	3.922
19	.688	1.729	2.093	2.539	2.861	3.883
20	.687	1.725	2.086	2.528	2.845	3.850
21	.686	1.721	2.080	2.518	2.831	3.819
22	.686	1.717	2.074	2.508	2.819	3.792
23	.685	1.714	2.069	2.500	2.807	3.768
24	.685	1.711	2.064	2.492	2.797	3.745
25	.684	1.708	2.060	2.485	2.787	3.725
26	.684	1.706	2.056	2.479	2.779	3.707
27	.684	1.703	2.052	2.473	2.771	3.690
28	.683	1.701	2.048	2.467	2.763	3.674

(Continues)

Critical Values for the t-Distribution (*Continued*)

29	.683	1.699	2.045	2.462	2.756	3.659
30	.683	1.697	2.042	2.457	2.750	3.646
40	.681	1.684	2.021	2.423	2.704	3.551
60	.679	1.671	2.000	2.390	2.660	3.460
120	.677	1.658	1.980	2.358	2.617	3.373
∞	.674	1.645	1.960	2.326	2.576	3.291

Index

A

Alpha, 36, 38–39
Analysis of variance (ANOVA), 41–42
    See also One-way ANOVA; Three-way
    ANOVA; Two-way ANOVA
Array, 180, 181

B

Bar chart, 10
Beta, 36
Bimodal distributions, 14
Bonferroni, Carlo Emilio, 69
Bonferroni test, 64, 69–71

C

Carry over effects, 135–136
Central limit theorem (CLT), 14–15
    to confidence intervals, 26–27
Central tendency
    mean, 3
    median, 3–4
    mode, 4
CLT. *See* Central limit theorem (CLT)
Coefficient of determination, 176
Confidence intervals, 3, 23–27, 178–179
Conservative, 55
Correction factor (CF), 46
Correlation
    assumptions of, 180–182
    and causality, 180
    computation, 173
    computational formula, 175
    confidence interval, 178–179
    heuristic and computational formulas, 187–188
    heuristic formula, 174–175
    hypothesis testing, 183
    independent correlations, 183–184

linear regression, 184–186
    point-biserial correlation, 193–196
    restriction of range, 182–183
    ρ, estimate of, 179–180
    sampling distribution, 176, 177
    testing the significance, 175–177
Correlation coefficient, 175, 179, 180
Critical values for Q, 201

D

Dependent correlations, 183–184
Dependent variable (DV), 34
Descriptive statistics, 3
    central limit theorem (CLT), 14–15
    central tendency, measures of, 3–4
    confidence intervals, 23–26
    dispersion, measures of, 4–9
    distributions, types of, 9–14
    z-scores, 16–23
Dispersion, measures of
    range, 4–9
    standard deviation, 8–9
    variance, 4–8
Distributions
    bimodal distributions, 14
    frequency polygon, 9–10
    leptokurtic distribution, 11
    negatively skewed distribution, 13
    normal distribution, 11–12
    platykurtic distribution, 12
    positively skewed distribution,
        12–13
    types of, 9–14
Duncan's multiple range test, 96–97
DV. *See* Dependent variable (DV)

E

Experimentwise error, 77–79, 96